JOHN GLOAG ON INDUSTRIAL DESIGN

Volume 6

HOUSE OUT OF FACTORY

HOUSE OUT OF FACTORY

JOHN GLOAG
and
GREY WORNUM

LONDON AND NEW YORK

First published in 1946 by George Allen & Unwin Ltd.

This edition first published in 2023
by Routledge
4 Park Square, Milton Park, Abingdon, Oxon OX14 4RN

and by Routledge
605 Third Avenue, New York, NY 10158

Routledge is an imprint of the Taylor & Francis Group, an informa business

© 1946 John Gloag

All rights reserved. No part of this book may be reprinted or reproduced or utilised in any form or by any electronic, mechanical, or other means, now known or hereafter invented, including photocopying and recording, or in any information storage or retrieval system, without permission in writing from the publishers.

Trademark notice: Product or corporate names may be trademarks or registered trademarks, and are used only for identification and explanation without intent to infringe.

British Library Cataloguing in Publication Data
A catalogue record for this book is available from the British Library

ISBN: 978-1-032-36309-7 (Set)
ISBN: 978-1-032-36649-4 (Volume 6) (hbk)
ISBN: 978-1-032-36652-4 (Volume 6) (pbk)
ISBN: 978-1-003-33310-4 (Volume 6) (ebk)

DOI: 10.1201/9781003333104

Publisher's Note
The publisher has gone to great lengths to ensure the quality of this reprint but points out that some imperfections in the original copies may be apparent.

Disclaimer
The publisher has made every effort to trace copyright holders and would welcome correspondence from those they have been unable to trace.

HOUSE
OUT OF FACTORY

by
JOHN GLOAG, Hon. A.R.I.B.A.
and
GREY WORNUM, F.R.I.B.A.

First Published 1946
All rights reserved

Dedicated to the men and women of Britain who have served their country in War, and rightly expect a home for themselves and their children in Peace.

Printed in Great Britain by
Bradford & Dickens, London, W.C.1

CONTENTS

	Page
Illustrations in the Text	vi
List of Plates	vii-viii
Advice to the Reader	ix

Chapter
1	Why the Factory-made House?	1
2	What Materials are Suitable?	8
3	What Methods Can be Used?	17
4	Can Homes be Mass-Produced?	79
5	How Long Will it Last?	86
6	How Much Will it Cost?	91
7	Will it be Comfortable?	102
8	Will it be Easy to Run?	109
9	What Will it Look Like?	132
10	Consider Your Verdict	136
	Index	139
	Plates	145

ILLUSTRATIONS IN THE TEXT

	Page
The Home-coming	x
American " bubble houses "	18, 19
Constructional details of the Weir house	21
Buckminster Fuller emergency steel housing unit	22
Constructional details of the Telford house	24
Plans of first prototype houses on Keyhouse Unibuilt system	26
Cross-sections of Keyhouse Unibuilt prototype houses	28, 29
Constructional details of second prototype Keyhouse Unibuilt house	30
Type plans of the Braithwaite house	31, 32
Constructional details of first prototypes of the Braithwaite house	33
General system of prototype Braithwaite houses	34
Constructional details of concrete clad houses at Northolt	36 to 40
Plan of Mark II Arcon demountable house	41
Plan and system of construction of Mark IV Arcon demountable house	42, 43
Plans of prototype houses designed for the British Iron and Steel Federation	45 to 47
Diagram of pair of Type A Howard houses	48
Plans and constructional details of Orlit concrete houses	49, 50
Constructional details of Scano I house	54
Constructional details and plans of Jicwood prototype house	57 to 60
Perspective and plans of Tennessee Valley Authority demountable multi-unit houses	61, 62
Plans and elevations of two-bedroomed type Tennessee Valley Authority houses	63, 64
Perspective and plans of trailer week-end house in aluminium	67 to 69
Mast-hung Dymaxion house	71
Plans for two-storey Uni-Seco house	72, 73
Standard Uni-Seco parts	74
Plans and perspective view of holiday house at Northport, Long Island	76, 77
Pierce Foundation experimental house	96
City of Coventry kitchen and utility room units	112, 113
Plumbing unit for two-storey house, by Dent & Hellyer	114
The Denham plumbing and heating system unit	116
Plan of " package " bathroom unit	117
Kitchen working unit recommended by Scottish Housing Advisory Committee	121
The Elcock house service unit	126, 127
Package kitchen-bathroom units	128, 129

LIST OF PLATES

Plate	
1	Timber houses in Carolina, U.S.A. and Sweden
2	Terrace of timber houses outside Stockholm
3	Swedish pre-fabricated timber houses
4	The Atholl houses and the Telford steel house
5	Weir houses of the Douglas type, and houses erected on the Atholl principle
6	Plywood house produced for the Fort Wayne Housing Authority
7	Houses in plywood produced for the American Dwellings Corporation
8	Pre-fabricated concrete wall in course of erection and house designed for the Celotex Corporation
9, 10 and 11	Tennessee Valley Authority sectional houses
12	General Housing Corporation multi-units
13	Demountable cottages on the Seco Unit system
14	Prototype house produced by Clothed Concrete Constructions, Ltd., on the Dyke system
15	Orlit concrete houses
16	Concrete clad houses by Hills Patent Glazing Company, Ltd.
17 to 22	City of Birmingham experimental houses with constructional details
23	Mark II Arcon house
24, 25	Details and interiors of Mark IV Arcon demountable house
26, 27	Type A Howard houses, exteriors and interiors
28	Two-storey plumbing stack of Type A Howard house
29 to 31	Third prototype houses produced by the British Iron and Steel Federation, exteriors and interiors

LIST OF PLATES *continued*

Plate

32 to 35	Models, layout and erection progress details of Keyhouse Unibuilt houses at Coventry
36 to 39	Braithwaite prototype houses at Hendon
40, 41	Exterior and interior views of Jicwood stressed skin bungalow
42	Plumbing unit designed and made by Grundy (Teddington), Ltd.
43 to 45	Model and plans of timber house designed by John P. Tingay
46, 47	Gas package kitchen and bathroom produced by Littlewoods, Ltd.
48	Denham pre-fabricated plumbing unit

Advice to the Reader

This book is not concerned with architecture. It is concerned with explaining to ordinary, intelligent people, how a great, new and energetic industry could quickly provide all the houses we want in Britain. If enough people would forget their pre-conceived ideas about architecture and building and make up their minds to live in their own stimulating, convenient and revolutionary century, all the people could have good, cheap, comfortable houses, so easy to run that harassed housewives would not be old women at forty—they would be young women still. For a number of reasons, variously ridiculous and insufficient, people once opposed the use of anæsthetics. When they were adopted an immense amount of pain and suffering was banished from life. We could banish drudgery *now* by using industry to make the homes that people want—homes that would be warmer, brighter, cleaner, quieter, and more private than those hitherto provided by traditional building methods.

Homes, made in the factory, are the product not of *architectural* but of *industrial* design. We have forgotten, hundreds of generations ago, the arboreal nest, the cave, the skin tent and the mud hut. Now we could forget old architectural fashions and all the cramping limitations they impose on home life.

THE HOME-COMING
(Pre-fabricated houses are now in the air)
(*Reproduced by permission of the Proprietors of* Punch, *from the issue of Feb.* 16, 1944)

CHAPTER I

Why the Factory-made House?

BRITAIN needs houses, and needs them quickly, and that need will long remain—a legacy of the waste and destruction caused by the second world war. Three factors have created this urgent necessity for an immense number of new homes :—
1. The need to get rid of all slums, not only in cities and towns, but in country villages.
2. The need to replace houses damaged by air raids, flying bombs and rocket bombs.
3. The need to provide homes for newly married couples.

Slums are composed of worn-out, overcrowded houses. Although Britain led Europe and America in slum-clearance schemes between the two world wars, thousands of slums remain. People robbed of their homes by enemy action must have new homes, so must young couples who represent the future health, wealth and happiness of the country. We must build as we have never built before, and with speed, economy and efficiency.

The vast demand for houses may be met by using the same industrial technique that enables manufacturers of motor cars and other mass-produced commodities to cater for large and continuously expanding markets. Many people believe that the ideal solution of this great housing problem is to use traditional building methods; but few would contend that the size and output of the building industry are equal to the task, save over a long period of years. The industry will have to cope with other demands, such as the repairing and rebuilding of our cities and the provision of new schools and other community buildings.

During the second world war the public was led to expect a high standard of labour-saving equipment in the post-war house. Official, semi-official and political propaganda about domestic amenities consistently nourished such expectations. They can be met only if such equipment is mass-produced in factories.

Factory methods call for two special qualities in building : (a) *precision methods,* in order to ensure the accurate fitting of equipment made in standard sizes, and (b) *dry methods of building*, to permit such equipment to be installed without risk of corrosion or deterioration. Factory-produced houses opened for public inspection during the war period vividly demonstrated the importance of this equipment problem. Few members of the general public have even questioned the methods of house construction ; their whole interest has been claimed by kitchen equipment—the sink, the laundry facilities, the provision of hot water, and storage for food and utensils. The bathroom has come next in importance, then bedroom cupboards, and finally, accommodation for pram, cycle and garden tools. Whether the roof of a house is pitched or flat, or whether the outside walls are of brick, steel or asbestos has seldom caused enquiry or comment. Although this disregard of constructional methods is characteristic of the home-loving British public ; although as individuals they give far more critical attention to the construction and mechanical efficiency of their bicycles, motor-bikes and cars than they do to the structure and planning of their houses, a term has come into current use that has aroused suspicion and encouraged prejudice. It is the term *prefabrication* ; and as any examination of the character, advantages and limitations of factory-produced houses is impossible without a reasonable appreciation of what that term implies, its significance will be discussed here.

Prefabrication is given a variety of meanings, according to enthusiasm, antagonism, or technical knowledge. For the purpose of this book it is taken to mean the preparation away from the site of the framework, walls, floors, roof and equipment of a house, so that such material may be delivered either from one or many sources, and assembled on prepared foundations.

For centuries it has been common building practice to prepare such units as window and door frames, doors, windows, stairs, cupboards and wall panelling in a joiner's shop, or , latterly, in some woodworking factory ; window frames and doors often being imported from countries where timber was abundant. In mediæval times roof members were made in workshops, transported to the site, and fixed in position by workmen when a building had reached the stage when beams, roof trusses and so forth, were needed. This idea of prefabrication has been widely accepted in the past, particularly in timber-growing countries, and its most comprehensive application has generally been accompanied by the use of timber not only for the framework of a house, but for the walls and roof. In Scandinavia, an architecture of wood had already developed while England was still only a collection of disunited Saxon states ; there, skill in woodworking was originally encouraged by shipbuilding. Scandinavian shipwrights attained a mastery over wood long before it was considered important to give much time or labour to increasing the comfort of houses. Gradually the practice of making independent units, whether for ship or house building, became established. It was a practice extended by the use of timber as a building material. Apart from Scandinavia, other European countries evolved an indigenous architecture of wood, and from the twelfth century to the seventeenth, English houses were timber-framed, and were really stout cages formed by oak beams, roughly worked to a square section with the adze, with the open spaces filled in by plaster, or, after the fifteenth century, by brick. Those roughly hewn timber bones of the house were first shaped not far from the woods where they were felled; they were then seasoned and ultimately carted to the site. English, Scandinavian and Dutch settlers on the North American coasts during the seventeenth century built their dwellings almost entirely of wood, generally unseasoned, and at first in the form of log cabins.

In England, after the half-timbered or wood-framed house with plaster or brick filling had ceased to represent common building practice, wooden houses were often built around a brick core, which contained the fireplaces and flues. These wooden houses, of which many were built between the middle of the seventeenth and the beginning of the nineteenth centuries, were framed and lined with wood, and externally clad with overlapping boards, known as weatherboards (or sidings) in England, and clap-boards in America. All the parts of these houses—doors, sash-windows, weather-boarding, framework—were prefabricated and stored in builders' yards, ready for use; just as bricks, tiles and other building units were stored ready for assembly. The frame house became a traditional form in the American colonies; it has endured to this day, and there is little difference in appearance, and none in the building principle, between the frame house that any lumber dealer in almost any State of the Union keeps in stock in parts, ready for immediate assembly, and those erected in New England, Virginia and the Carolinas during the Colonial period. Those white painted houses with their sash windows, shingled roofs, and graceful classic porches and front doors, had all the elegance and beauty that the Georgian period in England achieved in brick and stone; but the wooden American houses were built at greater speed and lower cost.

Prefabricated wooden houses were not unknown in eighteenth-century England. When Boswell mentioned to Dr. Johnson, his scheme for buying the island of St. Kilda, the Doctor said : " Pray do, Sir. We shall go and pass a winter among the blasts there . . . We must build a tolerable house : but we may carry with us a wooden house ready made, and requiring nothing but to be put up."[1]

But in England we have almost forgotten timber building and the large-scale opportunities it afforded for preparing material away from the site. Cheap bricks and the demands of the Navy during the Napoleonic wars stopped the building of timber houses. In America the practice persisted, which is one of the reasons why the idea of prefabrication finds readier acceptance in the United States now than in Britain, where it is regarded with suspicion as something new and disruptive, calculated to destroy all serenity, individuality and beauty in building. As a principle it is old ; as a practice it should be revived ; and although in England in the past and in America up to date, the fabrication of building units for assembly on the site has depended largely on the use of timber, there are other materials which could be used to produce good-looking, comfortable and *permanent* houses by methods which have been invented and perfected in factories. These materials and methods will be described in the next two chapters.

Most factory-made houses provide clear floor spans with no load-bearing partitions, and can, if and when necessary, be changed and re-arranged at small expense. On analysis, the traditionally built house compares unfavourably with factory-made construction on the economic grounds of transport alone. It has been stated that a normal six-roomed house weighs about 125 tons, six tons of this being water which has gradually to dry out ; and it has been calculated that it contains some fifty-three thousand separate parts. (The Lancaster bomber is said to contain somewhere about ten thousand.) Twenty thousand of these separate parts are bricks, and each brick has to be manhandled several times in manufacture, transport and fixing.[2] Such a house has to carry a load of about six tons, represented by the inhabitants and their belongings. The pair of " Unibuilt " experimental houses erected at Coventry during 1944 had a total weight of 51 tons between them. A later development of this system at Edinburgh, with a lighter external cladding unit, weighed considerably less. The prototype design of a single storey house in aluminium alloy erected by the Aircraft Industries Research on Housing, weighed 1.9 tons. Apart from weight, which affects transport and labour-handling costs, traditional building demands site-labour operations,

---

[1] Boswell's *Life* : second edition, Vol. II, p. 4 (1772).
[2] These facts and figures are given by Mr. H. V. Boughton, A.M.I.C.E., in *Building*, October, 1943.

such as planing, hammering and sawing, repeated thousands of times. Some twenty independent trades are needed on the job, some for a few hours only, involving many sub-contractors. Even with good organization, bottlenecks are bound to occur, and hours of time are wasted. Can house construction under such conditions be considered economically efficient?

The underlying principle of the factory-made house is to prepare by mass production as many parts and units of the house as possible, thus reducing erection time and site work to the minimum. Such parts and units made under factory conditions of controlled power and mechanical operations achieve accuracy, efficiency and economy. The ideal of prefabrication is to increase the size of the factory-made units to a maximum, until the whole side of a house or a complete room is produced. This ideal has been attained under wartime building conditions in America. Increased sizes of wallboard covering, such as Homasote, and greatly improved adhesives have helped to make this possible. But it is obvious that the larger the mass-produced unit, the greater is the tendency towards monotonous standardisation in the completed house. In Britain, designers of prefabrication systems are conscious of this danger, and the size of wall unit selected for most of their systems is within a maximum 4' width, generally less, and of a one or two storey height, thus allowing considerable flexibility in planning and composition, so that individual character and taste, for the preservation of which we have fought two world wars, may be adequately accommodated. Traditional building has by no means avoided monotony in the past. Council housing schemes, however well sited, have been built in pairs and blocks to a monotonous pattern. The working-class tenant has had practically no say regarding his tastes or requirements and has come into the category of an institutional inmate, being denied the right to live as a free citizen in the home of his choice. He has been provided with the barest minimum of equipment and, as a weekly tenant, he has had no incentive to increase or improve this at his own cost. While the factory-made house may be varied in plan and composition by the skilful arrangement of its units, nobody is likely to complain of its standardised uniformity of equipment. For instance, an assembly of plumbing units, produced in the factory by the thousand, can give far better value for money and reach far higher efficiency than anything installed *in situ* in traditional building. Space heating and water-heating units can likewise be better produced with their flue pipes designed as part of the unit. Kitchen fitments would be manufactured as separate units, which could be arranged to accord with variations in plumbing. All such installations and equipment in a factory-made house may be easily and inexpensively renewed at future times, to meet wear and tear, or to allow advantage to be taken of new and improved appliances, without pulling out any of the structural parts—an extremely important consideration.

Another advantage of the factory-made house over the traditionally built types is the greatly improved insulation against heat, damp and cold that becomes easily and cheaply available. Effective insulation means the difference between comfortable and uncomfortable living. The economic and efficient warming of the home in this climate is of the highest importance. *It may not be generally known that it is possible, through a 9" thick brick wall, to blow out a candle with a bicycle pump.* Apart from outside temperature, wind in this country is a big cause of heat loss. The earlier practitioners of prefabrication, in the days of the " All Steel " and the " All Concrete " house, neither knew much about insulation, nor gave it much attention. In recent years, considerable development has taken place in the production of insulating materials and the embodiment of these is considered an essential part of prefabricated house design. Such materials as aluminium foil, wood wool slab, rock wool, glass wool, gypsum plaster and asbestos, are made in various forms, block or sheet. It is a relatively simple matter to incorporate such materials in a factory-made house.

In speed of erection the factory-made house finds no competition from traditional methods. Closely connected with this question of speedy erection is the immense advantage of any system of *dry* building, the most important aim being to get the roof on as soon as possible. An instructive attempt to achieve this in traditional building was made before the war by the British Steelwork Association in collaboration with Mr. Dennis Poulton, A.R.I.B.A. A steel framework for the house was quickly erected and a fully completed tile roof then put on top. With the first floor framework serving as a platform, building proceeded on two levels at once under cover.

Such construction cannot be considered sufficiently economical for mass-produced housing. The light gauge metal sections constituting the prefabricated frames of to-day provide a new form. One system, for example, claims that the light steel chassis can be erected on the site for one house by four semi-skilled men in six hours, the roof slabs and covering can then be put on immediately.

It has not yet been shown in America or elsewhere that the factory-made house can be produced more cheaply than one traditionally built. But comparisons between the equipment and service installations in each must be taken seriously into account. *No council house erected before the war can offer anything like the amenities or equipment provided in the Portal experimental house.* From our industrial experience, whether in the production of motor cars, refrigerators, or any other mass-produced article, we know that if a sufficient market to justify large numbers exists and is maintained, production costs are correspondingly reduced. The more a manufactured unit can be repeated in any one house, the less dependent is the manufacturer upon orders for large numbers of houses at a time. Encouragement is thus given to designers of prefabricated houses to restrict their units to sizes that allow the greatest flexibility and ultimate variety in the finished house.

After the first world war, new techniques in building in Britain were stimulated by a great shortage of housing and skilled craftsmen. Experiments were made in steel, timber and concrete. A brief survey of the results of such experiments, both in this country and elsewhere, is given in the next two chapters.

CHAPTER 2

What Materials are Suitable?

MANY materials may be conveniently used in the mass production of units for the factory-built house. Metals such as steel, cast iron, and aluminium and its alloys are available ; also composite and chemically produced substances, such as concrete and plastics ; while timber in all its forms—including plywood impregnated with synthetic resin or faced with plastic or metal sheet—is the traditional, basic material for prefabricated buildings. From these materials specific types of factory-made houses have arisen ; many as a result of some pressing need to use a particular metal, such as steel, or of some obvious economic advantage to be derived from an indigenous and abundant material, such as timber in Scandinavia or North America. Before the nineteen thirties, the form and constructional character of most prefabricated buildings were generally dominated by the properties and the limitations of the chief material employed. They were " all wood " or " all steel " ; and very often the chief concern of their designers was to disguise them so that they were indistinguishable from houses built by traditional methods. Comparatively few experiments were based on the idea of producing a cheap, convenient dwelling by selecting the most appropriate and easily assembled materials. The contribution of the principal prefabricating materials to the development of the factory-made house should be briefly examined.

STEEL

Between the two world wars, some thirteen thousand steel houses were erected in this country : ten thousand in England and three thousand in Scotland. In appearance, these houses gave little indication to the public of their construction, for they had pitched tile roofs, standard windows, chimneys and smooth walls, like millions of other semi-detached dwellings. The chief incentive for this form of construction came from Scotland. Lord Weir and the late Duke of Atholl made a serious attempt to relieve the severe unemployment in the Clydebank shipyards during the nineteen twenties, by finding a new outlet for steel construction in housing.

Political opposition from the building Trade Unions was the main cause of stopping steel house development. During the acute shortage of skilled labour it was sometimes necessary to employ non-building trade labour for prefabricated housing, both in the factory and on the site. With the Weir houses this caused serious disputes, as engineering Union labour was employed and paid at lower rates than unskilled building trade labour.

In the U.S.A., prefabrication in steel was stimulated by depression in the industry during 1930 to 1933. Unfortunately, energetic attempts to find an outlet for surplus material resulted in the production of " All Steel " houses, and ingenuity was concentrated upon finding the maximum rather than the most appropriate use for this material.

This American work was contemporary with extensive German development in steel houses, and was anticipated by the British experiments. Until about 1930, American designers appeared to have no inkling of the experimental uses of steel in other countries.

German steel systems arose from British experience, gained in the early nineteen twenties, especially from the Weir house. German development began about 1926, when the building of steel houses in Britain had almost ceased. It was stimulated by the depression in Germany's steel industry, and it coincided with the launching of big housing projects. It was rapid and varied, and was still evolving and disclosing great ingenuity when in 1933 it was stopped by the Nazi accession to power and the withdrawal of steel from the housing market.

Lack of continuity in development of the steel house in Britain has checked any progress comparable with, say the design of the refrigerator, motor car, or long-distance coach. But the second world war has brought some compensations for this loss. Many new varieties of steel and of light metal alloys have been evolved. Production of strip steel has been greatly advanced, and the cold rolling and pressing processes highly developed. Welding technique has also been greatly improved. Above all, industries have begun to recognise the value of composite production, which allows each part of a manufactured product to be of the material best suited to its purpose. Thus in recent examples of new systems there is a tendency to combine light-gauge steel with timber for structural members, whether for walls floors or roofs. The timber has particular value in offering a fixing surface for nails or screws, and can reduce sound transmission. The metal parts save bulk of timber, which is vital while timber is in short supply, and also save weight.

Steel construction in houses is obviously suited to factory production. Precision-accuracy is required, and with such material, adjustment on the site is costly and difficult. By combination with many materials now available, very high insulation can be obtained from heat, cold and damp, with minimum weight for transport and handling. Sound insulation with welded construction is a special problem for the designer, but successful solutions have been made. Protection of the steel from corrosion may now be carried out in many ways that would have been difficult to achieve before 1939.

CONCRETE

Generally, the special characteristics of concrete, its weight, mass, brittleness and coarseness, do not lend themselves very readily to large-scale prefabrication. But concrete does lend itself to the casting (or prefabrication) on the site of component parts such as beams, stanchions, panels and blocks. Moulding processes for such components must be kept simple and by the establishment of a " site factory " transport reduced to a minimum. Difficulties can arise because of the large space needed for the moulding and curing processes, and the capital value of the moulds themselves. For economy, these moulds must each be used many hundreds of times, implying large-scale contracts and simplification in number and type of element.

For economic handling, simplification of design and plan and standardisation of structure are vital. In the best examples, where large scale mechanical handling has been introduced, the site layout has been in long simple rows, allowing a continuous runway for the crane. In such cases, it is not possible to consider this site layout as something separate from the constructional problem.

Restrictions in the use of concrete in the past have been due to weight, weather resistance and lack of sufficient insulation properties.

The development to-day in lightweight aggregate concrete has done much to remove these restrictions. Such aggregate can be (1) natural (such as pumice) or (2) in the form of by-products (such as clinker or breeze) or (3) artificial or processed.

The first alternative is not available in this country, and the second has to be used with caution, since it may contain dangerous constituents. The third offers foam slag as a suitable processed aggregate for immediate use in this country for either moulding or *in situ* pouring. Other processing can take the form of oven heating to turn the moisture content into steam, so eventually creating a kind of pumice, or by foaming through chemical means.

Some types of light aggregate lack the strength of " dense " concrete, but additional strength can be provided by using a proportion of sand in place of the fine gradings.

TIMBER

Although timber building for houses ceased to be common practice in the early years of the nineteenth century, sectional timber buildings have for many years been marketed in Britain. Their use has been more generally for sheds and out-buildings than for dwelling houses. Few developments in prefabricated timber construction have occurred in the inter-war period and in wartime, timber for building purposes was severely restricted. Nevertheless, between 1937 and 1941, a total of 2,246 timber houses were built in Scotland. The development of plywood construction in Britain for wartime purposes, for ships and aeroplanes, is likely to encourage timber fabrication in factories in the future. We are compelled to import most of the timber we need, and our economic circumstances will demand a much more scientific use of the material than in the past. This is already beginning to show itself in several types of house construction, where fabricated plywood beams are built up in preference to the use of solid timber.

In the U.S.A. prefabricated timber houses have reached a development and output far ahead of any other country. The idea of prefabrication already familiar was extended, and its practical application increased a hundred years ago by the invention of the machine-cut nail and the power saw, which between them finally transformed the log hut into the wood frame house. The making of such houses was originally confined to the preparation of scheduled material, the erection still remaining a handicraft process.

The year of great trade depression, 1929, inspired interest in the potential market offered by low-cost housing. Idle factory space and the collapse of normal markets gave life to a new industry, and house prefabrication received energetic publicity. Beyond publicity, the ten years that followed witnessed no spectacular results ; but during that time a background of research and experiment was created which made possible the great feat of building some million and three-quarter wartime houses in two years. In that ten-year period, the development of large sheet material took place, such as Celotex, Homosote and Douglas Fir plywood, which, together with plastic glues, permitted large panel construction. In prefabrication, the larger the panel the less the site work, provided weight does not become too great for handling. Also, the larger the panel, the less the number of visible and vulnerable joints.

Accompanying a more scientific use of timber to-day, is the campaign to establish an internationally recognised grading of the material. Until now the absence of such standards has demanded a most extravagant factor of safety for structural timber members. That a sound timber is able to do probably twice the work of a faulty one has received little recognition. Great advances in the scientific seasoning of timber have been made in recent years, and the close control of moisture content now possible, while making for greater efficiency, favours any system of dry building and minimum of site exposure. Wartime experience in the impregnation of timber with plastic resins has revealed new uses for the material, enhancing its weather- and fire-resisting properties. In addition, timber hitherto considered unusable or mere waste is now fabricated into wallboards and hardboard. Resin impregnation of such boards again enlarges the range of valuable material for building.

PLASTICS

Chemically produced substances, which may be shaped by the application of heat and pressure, obviously facilitate factory production, if they are used, wholly or in part for prefabricated building units. Such substances are known to-day as plastics, and their influence upon the design of innumerable commodities has already been great, and is likely to be still greater in the future.[1] But so far, plastics have not played any conspicuous part in building. They have been used for various fittings in the home, but for little else. This has been attributed to the following reasons:

1. The high initial cost of material compared with other raw materials.
2. Processes that are suitable for mass production involve high tool costs for various mouldings.

---

[1] The subject is discussed in detail in *Plastics and Industrial Design*, by John Gloag. (George Allen & Unwin, Ltd., 1945.)

WHAT MATERIALS ARE SUITABLE? 13

3. Only a few plastic fabricating firms have been organised to handle suitable products for the building trade.
4. Until recently, technical information regarding properties and standards has been difficult to obtain : without it architects and designers are unable to make adequate comparisons between plastics and other materials.

There are direct and indirect uses for plastics in the home. The indirect uses cover plastic compounds—paints, lacquers and adhesives. The direct uses include building components, such as sheets, tiles, cover strips and other mouldings, and an immense range of small fitments, such as door furniture, lighting fittings and the bathroom and kitchen equipment. These direct uses covered a large market before 1939. Further developments are likely to take place in the indirect uses, particularly in the application of resins to other materials such as sawdust and pulp, asbestos and plywood, for the production of wallboard and other purposes. Expanded plastic in cellular form has been much used during the war for insulation and may, in conjunction with other materials, have practical value for room partitions. The use of plastic beams and supports, whether solid or hollow, may prove too costly for adoption in building.

But the association of plastics with other materials promises developments of great potential value in the mass production of building units. The technique of using materials in association has been accelerated since 1939. It has been said that : " To any student of industrial design, it is obvious that many materials have complementary uses, and that productive partnerships between materials are not only possible, but almost inevitable. In the course of war production many such partnerships have been established between old and new materials, some of them resulting from wartime research work. For instance, aluminium has been welded to glass ; combinations of plastics and light alloys have been used ; plastic sheets and plywood have been cemented together ; plywood has also been impregnated with various synthetic resins, so the best of both materials is available. Resin-bonded plywood preserves much of the character of wood, and eliminates many of its disadvantages. These various partnerships, and the changes wrought in the traditional character of old materials, allow the designer not only to create objects that are perfectly adapted to perform specific functions out *to create the material* from which they are made."[1]

---

[1] *Plastics and Industrial Design*, by John Gloag, Chapter III, page 28.

LIGHT METAL ALLOYS

The light metals of industry are aluminium and its alloys, and magnesium, though the latter is unlikely to be used for building purposes to any appreciable extent. Aluminium is the most abundant metal on the earth's surface. We are told that the crust of the earth is 8% aluminium and 5% iron. Since 1937 aluminium production has increased enormously, stimulated by wartime requirements. This greater production capacity should not only increase future supplies, but reduce prices and bring into everyday use a metal that was previously almost in the luxury class. For certain uses, aluminium has many advantages over steel, by virtue of its light weight, its resistance to corrosion and its decorative appearance. In welding, processes have been greatly improved and the X-ray now assists the inspection of welds and testing.

Fresh developments in alloys and improvements in anodising the material can provide finishes which make possible the elimination of paint and consequently reduce upkeep. It has been estimated that the annual wastage in this country due to the rusting of iron is equivalent to 20% per annum. Heavy steel structures demand annual expenditure for painting. With light steel construction, special treatment is necessary, otherwise an increased factor of safety is required to counteract corrosion risks. In sheet form, aluminium can provide excellent roofing material, and serve also as wall cladding, internal or external. In extruded and fabricated form, it is well suited for windows and doors, as well as for many types of fitting in the house. As foil it has already proved itself an excellent thermal insulator.

CAST IRON

This is a traditional material, with many new forms and finishes, which is particularly well adapted to mass production. One of the most familiar examples of the use of cast iron for prefabricated structures is the G.P.O. telephone call box, designed by Sir Giles Gilbert Scott, O.M., R.A. Its capacity for resisting wear and tear is considerable, as the G.P.O. pillar box proves, for nearly every letter posted in the street is protectively surrounded by cast iron. This was the material chosen for the structure of the greatest adventure in prefabricated building to date, namely, the Crystal Palace, designed by Joseph Paxton, for the Great Exhibition of 1851. It was made in various Birmingham workshops and was erected in Hyde Park for six months. Its floor area was four times that of St. Peter's, Rome, and it was sufficiently demountable to be re-erected finally at Sydenham.

An American architect, James Bogardonson, designed various buildings for the American World Fair of 1853, with façades consisting of repetitive units of cast iron and glass ; a system clearly inspired by the prefabricated units of the Crystal Palace.

Cast iron was used on a large scale for standardised units and supports for various kinds of buildings throughout the nineteenth century in Britain.[1] Attempts were made between the wars to use this material for housing units ; but limitations in casting technique involved the use of quite heavy sections. Sheets in pan form were most commonly found of 3/16ths of an inch thickness. Modern mechanical casting can produce these accurately to-day at under $\frac{1}{8}$th of an inch thickness. Such material was very sound, for it was not so subject to corrosion as steel, though not sufficiently economical to have a wide, practical use. Modern casting methods, and particularly enamel finishing, are likely to give cast iron a useful place in the factory-made house of the future, though for equipment rather than for construction.

SOME CONCLUSIONS ABOUT MATERIALS

In this highly condensed account of the principal prefabricating materials, the importance of partnership between materials has been emphasised, and the use of the most appropriate substances in combination for the mass production of house-building units. Attempts to use any material merely for the sake of using it, because the industry concerned with producing and fabricating it needs some large and immediate " outlet " to survive a period of economic depression, lead almost inevitably to the sacrifice of fitness in design and convenience in use. The development of the factory-built house would be retarded if the industries concerned with its production were hampered by the need to " use more " of this, that or the other material. The American " All Steel " house, mentioned on page 9, is an example of the deflection of design from its primary object of serving the ultimate consumer by using to the best possible advantage the most apposite and economical materials for the factory-made dwelling.

However, designers of prefabricated houses have outgrown the days when they thought in terms of the " All Steel," " All Concrete " or the " All Timber " house. They realise that the only satisfying answer to

---

[1] Accounts of the early architectural uses of cast iron are given in *John Nash*, by John Summerson (George Allen & Unwin, Ltd., 1935), Chap. II, pages, 43-46 ; in *The Missing Technician in Industrial Production*, by John Gloag (Allen & Unwin, Ltd., 1944), Chap. I, page 19 ; and in *The Lives of the Engineers*, by Samuel Smiles (John Murray, 1861), Vol. III. The most comprehensive book on the subject is *Cast Iron in Building*, by Richard Sheppard, F.R.I.B.A. (George Allen & Unwin, Ltd., 1945).

their systems is to be the " Composite." This provides a safeguard against monotony. For instance, the sponsors of a light steel frame system are able to adopt many alternatives for outside clothing. This makes possible a large variety of finishes and colour on any single scheme without interfering seriously with mass production methods. Several systems allow of an outer brick skin to be built, to accommodate the taste and secure the confidence of those who still cling to " Bricks and Mortar."

CHAPTER 3

What Methods Can be Used?

SYSTEMS of prefabrication have multiplied since the early 'thirties; new ideas and technical improvements will undoubtedly produce fresh systems and modifications and extensions of those now existing; and already the study of the subject is complex, and would be utterly bewildering unless it was examined in terms of *constructional principles*. Here we are chiefly concerned with systems that may be described as complete; though that term should be qualified, for few existing systems are or can be absolutely complete, because certain site work is inevitable. There are many *partial* systems—for example, those which entail the pouring *in situ* of concrete structural members, or wet *in situ* processes for making the inner or outer skins or the floors. Complete systems are largely based on the factory production of parts to facilitate dry processes of construction. But whether the systems are partial or complete, they may be classified according to their constructional principles under one of the following headings, easy enough to remember for each may be compressed into a single word :—

1. SHELL SYSTEMS : Houses with a weather protecting skin sufficiently strong to act as structural support.
2. SKELETON SYSTEMS : Houses with a weather protecting skin attached to a structural frame.

Traditional building comes under these headings : load-bearing brick or stone walls under the first, and timber or steel-framed buildings under the second. Prefabricated *Shell* systems vary from the assembly of large pre-cast concrete sections to the "airform," which is simply a balloon of fabric, blown up on the ground and then sprayed with concrete. (Houses of this type have been erected in a desert colony in the United States.) Many *Skeleton* systems rely considerably for strength and rigidity on the panel infillings between supports and beams, so that the skin becomes partially structural. Very many of the systems that have been tried out and adopted, either for large or

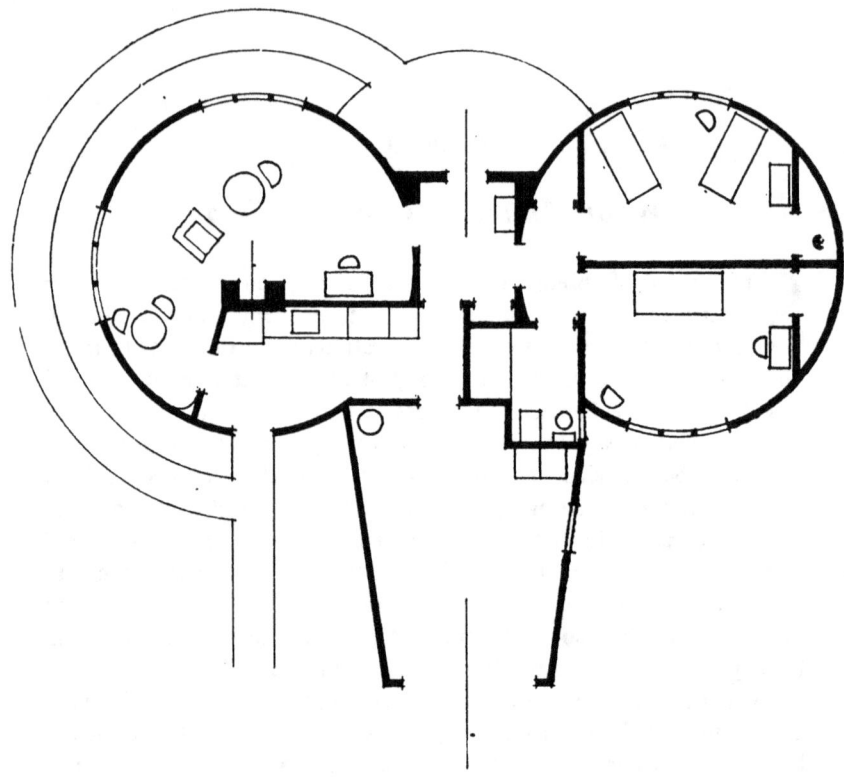

Type plan of a bubble unit. (See opposite page)

small scale experimental housing schemes, are *Shell*, while most of the early experiments in prefabrication are *Skeleton*. We shall review briefly various pioneer, existing and experimental systems in relation to specific materials. No attempt is made to give an exhaustive list of systems: there are hundreds of them, and there will be hundreds more before the end of this century; but those described show one or other of the two basic constructional principles in operation, and indicate how the inventive use of old or new materials or combinations of materials develop fresh forms. Most of the systems included in this chapter appear under *Steel*, *Concrete* or *Timber*: plastics and light alloys are largely contributory materials in composite systems. Some pioneer experiments in cast iron are mentioned. The main classifica-

" Bubble houses " in an American village. Architect: Wallace Neff. *Above:* A bubble unit. The living room is 14 ft. by 23 ft., the bedroom is 14 ft. by 10 ft.
Below: A single unit

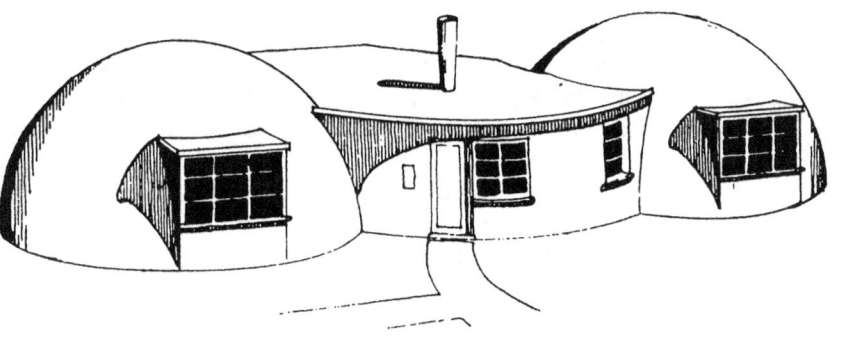

tion of every system is inset in the margin in a heavier face of type, so it is immediately clear which is **SHELL** and which is **SKELETON**.

SKELETON

STEEL

Experimental work on steel houses began in Scotland during the nineteen twenties. They had frames of wood and skins of steel. Most of the Scottish steel-built houses are owned and maintained by the Second Scottish National Housing Company (Housing Trust) Ltd. It was founded to finance some 2,557 steel houses on fifteen various sites, all designed and erected complete with streets and services under the supervision of Mr. A. H. Mottram, F.R.I.B.A. (1925 to 1928). It

SKELETON

should be emphasised here that some such substantial market was essential in order to make economically possible the mass production of the units used. The houses were mainly of three types, Weir, Atholl and Cowieson. While the load-bearing frame was of steel in the Atholl type, the other two types were timber framed. All types were clad with thin steel plates, with various internal finishings of wallboard or even plaster. When first introduced, their cost compared favourably with other houses being built at the time. By 1928, this advantage disappeared, and the production of steel houses generally ceased. A thorough examination of many of these houses has been made by the Building Research Station, and its findings have been published.[1] A " paint harl " finish of chipped granite was eventually found to be the most satisfactory for the outside walls, and in many cases although no redecoration was done between 1936 and 1944, no serious defect occurred. These houses have a high degree of thermal insulation, due to the precautions taken to prevent heat loss through the walls and to the favourable properties of the internal wall and ceiling coverings ; but sound insulation was not so successful, weakness in this respect occurring particularly through party walls. As homes they have proved very popular with tenants.

SKELETON

While several of the Atholl houses were built in England, various other forms were developed, particularly the Dorlonco type. These Dorlonco houses were constructed with pre-cut and numbered steel frames covered in expanded metal and sprayed from a cement gun. Many of these proved defective, due to insufficient cover of metal lathing which rusted and caused the cement rendering to crack and fall off. Generally speaking, the construction was not simple and it was closely associated with traditional building methods so far as foundations, services, roof covering and so forth were concerned, and did not represent a really comprehensive degree of prefabrication.

---

[1] *Post-War Building Studies*, No. 1 : *Housing Construction*. Published for the Ministry of Works by H.M.S.O., London, 1944. Paragraph 526, page 87.

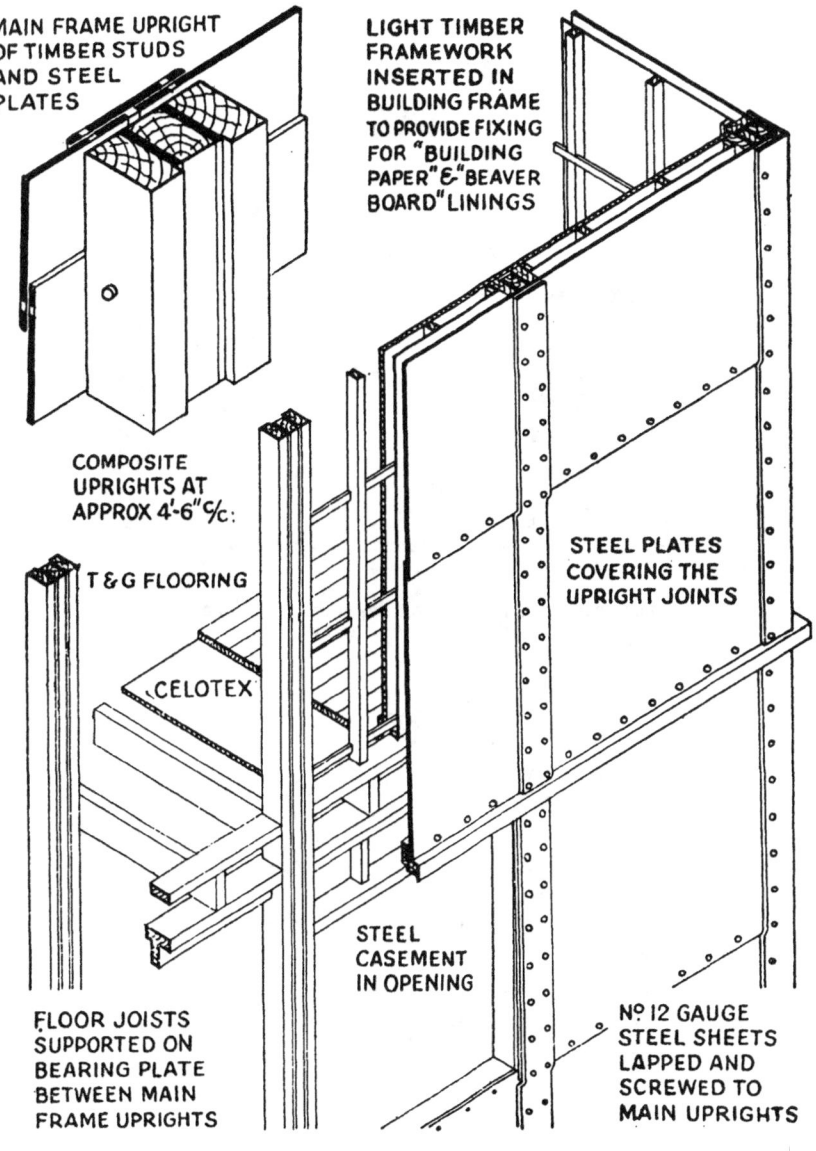

Constructional details of the Weir house. Drawing by George Fairweather, F.R.I.B.A.

(*Reproduced by courtesy of the* Architect and Building News)

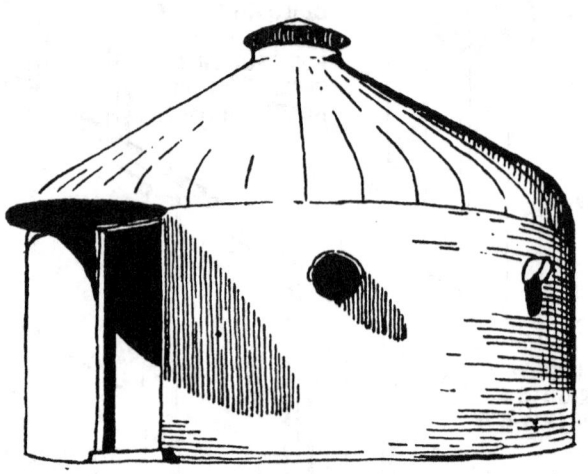
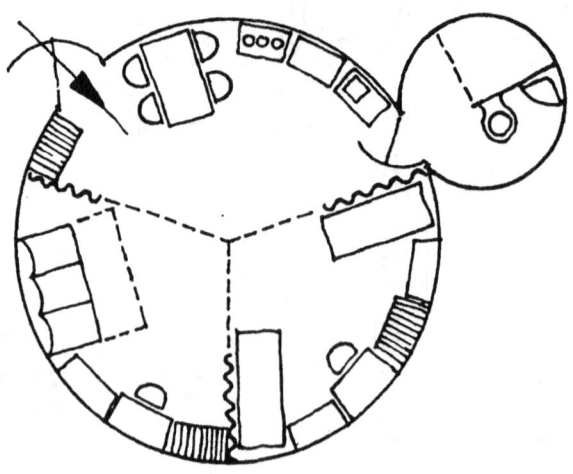

Emergency steel housing unit designed by R. Buckminster Fuller: elevation and plan

WHAT METHODS CAN BE USED? 23

SHELL   A system of flanged steel plates welded together has recently been produced by Lord Weir for single storey buildings. The sheet steel with its turned flanges is sufficiently strong to carry the roof, and delivered in large welded sections allows for a minimum of bolting and welding on the site. The system would be applicable to buildings with two storeys.

SHELL   The original Portal emergency house was an example of sheet steel skin that was given the necessary strength to carry the weight of its roof by pressing the sections of its panels.

SHELL   In the U.S.A. the circular houses designed by Buckminster Fuller for emergency use in large industrial areas also exemplify the use of a steel-supporting skin. They are designed in the form of dumpy steel cylinders 20′ in diameter, with horizontal ribbing, built up in sections and bolted together, with a flat, pie-shaped roof, also in sections, and porthole windows inserted in the walls. The steel used is 20-gauge and the inside is lined with wallboard and insulated, the internal divisions to rooms being by curtain only. Though their appearance suggests grain bins, they are original and practical in conception.

SKELETON Among early examples of American steel-built houses was the Nils Poulson house, built in 1890 with a close-spaced steel frame, brick infilling and copper sheathing. It was still in excellent condition when pulled down forty years later. Another was the Naugle house built in 1907. This also was a close-spaced steel frame, but it was a retrogressive type, for its construction was based on typical timber framing. No development of importance occurred until the mid-nineteen twenties, when a few houses were built that followed the Naugle method and used a metal frame that followed a timber prototype. In these examples traditional wood spacing was gradually abandoned, thus leading to economy in material. Rapid developments then

SHELL   followed of sheet steel load-bearing panels such as Armco and Steelox. This type of construction had been used in the form of trays since about 1925, and in the form of box panels or corrugated sheets after

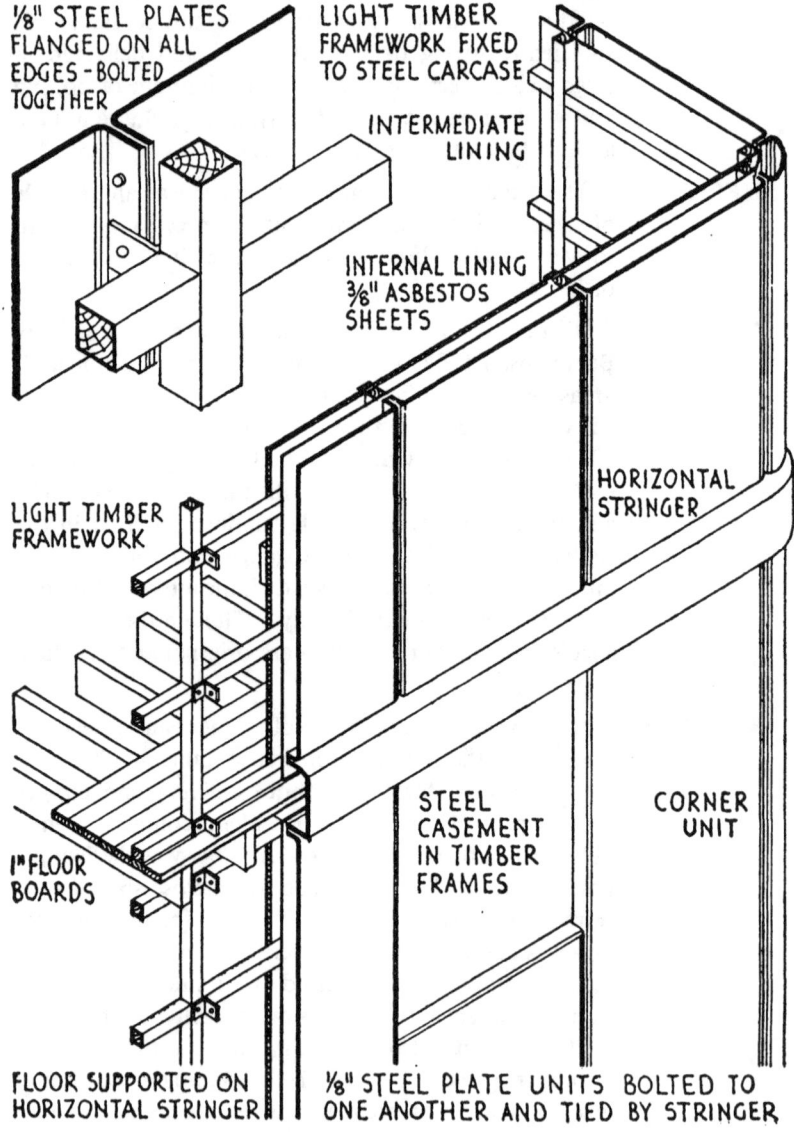

Construction of the Telford house, manufactured by Braithwaite & Co., Engineers, Limited, during the inter-war period. (See plate 4)

(*This drawing by George Fairweather, F.R.I.B.A., is reproduced by courtesy of Braithwaite & Co., Engineers, Limited, from the* Architect & Building News)

WHAT METHODS CAN BE USED ? 25

SKELETON 1930. Pressed steel in other forms was widely used, and there were a number of close-spaced frame and panel types. Another development, peculiar to America, was the production of pre-cut metal sections. Later in 1935, Stran Steel introduced their present " metal lumber " in pressed steel, and to-day a great range of lightweight metal sections, rolled or pressed, is on the market.

SHELL In Germany, the first steel house recorded was the Scherrer house brought out in 1915 by a firm which had previously produced prefabricated timber houses. It was a load-bearing panel type. The extent of its use is not known. The actual volume of steel house building in Germany after the first world war was not much greater than in Britain, but many more systems were

SKELETON evolved. Both close-spaced and open-spaced frame types were used. The former were either clad with sheet steel or composite panels, or were filled in with lightweight pre-cast concrete slabs. In one type a steel plate cladding was used which incorporated a key for stucco. Open-spaced frames were usually applied to multi-storey structures, and there was a considerable development of this method for three- and four-storey blocks of working-class flats. The frames were sometimes erected in large sections on the ground before being hoisted into position, anticipating the later methods of Mopin in

SHELL France. Great progress was made with load-bearing panels, which were usually of tray shape, one-storey high and about three feet wide. These panels were also used for the roof. Again this German method emanated from a British prototype—the earlier British

SHELL cast-iron tray systems and the Braithwaite house. The Germans experienced difficulty with steel, because of its poor insulation qualities and high ratio of expansion. Insulation was frequently incorporated as an integral part of the panel when it served to prevent condensation on the metal, special joints being evolved which allowed for expansion and contraction in the steel.

In France there is little to record prior to 1927. After that date, some systems were sponsored by the French

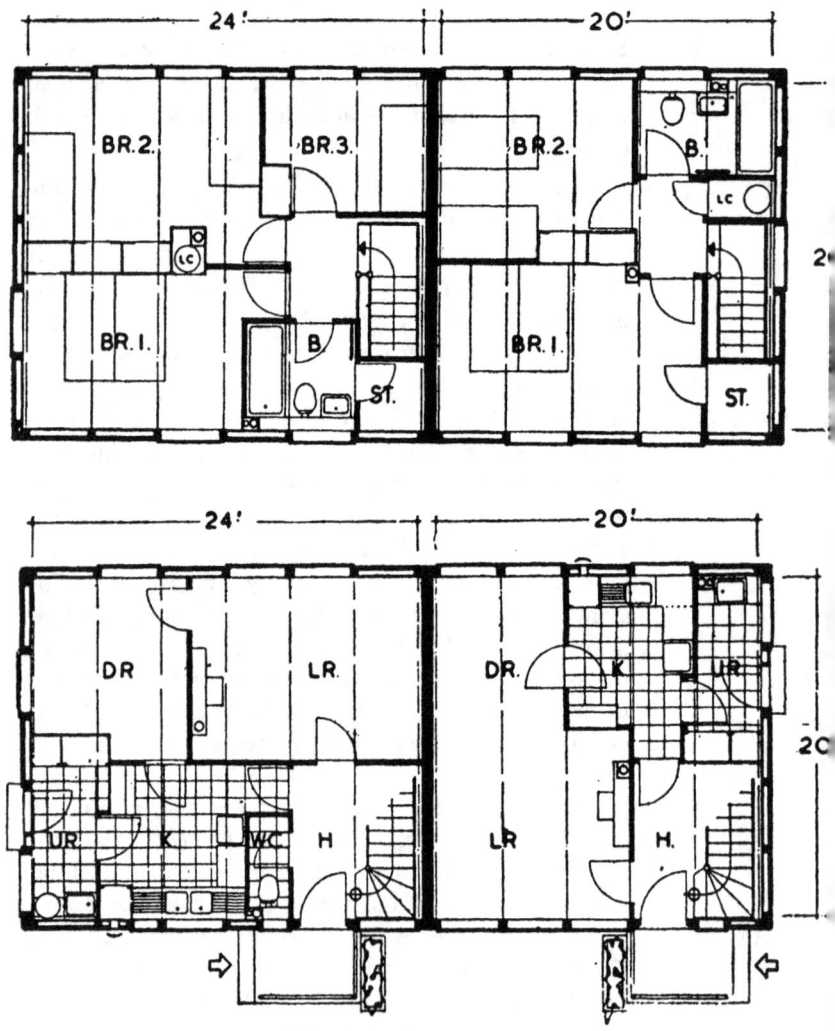

Ground and first floor plans of the first prototype houses erected for the City of Coventry on the Keyhouse Unibuilt system. (See plate 33.) Architects: Grey Wornum and Richard Sheppard, FF.R.I.B.A.

SHELL   steel industry. Single- and two-storeyed buildings were erected with internal and external wall faces, one system being similar to Robertson or Armco units, which were later so successful in America. In both countries these units were welded into large size walling sections, but were more widely applied as flooring units. Maximum economy was not obtained because plans were not adapted to suit a regular frame structure, but the constructional details were most carefully studied and no failures seem to have occurred. In 1928 there was a

SKELETON   housing exhibition in Paris and houses were built incorporating steel frame units, light enough to be handled by one man, and sheathed with asbestos or gypsum. In 1939 the Technical Bureau for Steel Utilisation in Paris sponsored a competition for flat construction, and sample sections of structure were erected. These included steel sheathing units prefabricated with integral insulation, load-bearing sheet-

SHELL   steel units, and frame units for the formation of service ducts.

During the second world war, a small but productive amount of design research work was conducted in Great Britain, and several new systems of prefabrication were evolved. Some of these had been carried to the building stage in 1944 and 1945. For example, at

SKELETON   Coventry and Edinburgh, the Keyhouse Unibuilt system was demonstrated by experimental pairs of houses.

This system consists of a chassis of lightly constructed panels 4′ wide and either 10′ or 20′ high, made of 18 gauge channel section $1\frac{1}{2}'' \times 1\frac{1}{2}''$ strip-steel projection, welded together. The floor and roof trusses are built up of similar section, and span from wall to wall. They bear on the junction of each wall panel at 4′ centres. The unusual depth of these trusses, 20″, ensures rigidity for the whole structure. Except for one fixing bolt at the end of each truss and for small bolt fixings to the foundations, the joining of all wall sections is by simply designed inter-locking parts, easily demountable.

It is claimed that four semi-skilled men can erect this structure on its foundations in six hours, and would require neither plumb bob nor level to do so. Outside clothing units of wood wool, faced with asbestos, or pressed metal panels lined with glass wool or other insulating material, are fixed by clips sliding into the metal channels of the chassis. This outer skin provides external insulation to the steel structure and takes up any movement in the frame caused by temperature changes. The clothing units are jointed on their outer faces with special mastic inserted by a pressure gun. The walls are lined inside with plasterboard, $\frac{1}{2}''$ thick, and this, with cavity construction, provides thermal insulation equivalent to that of a 22" thick brick wall, while furnishing an inner skin that will warm up in a room far more quickly than a solid wall.

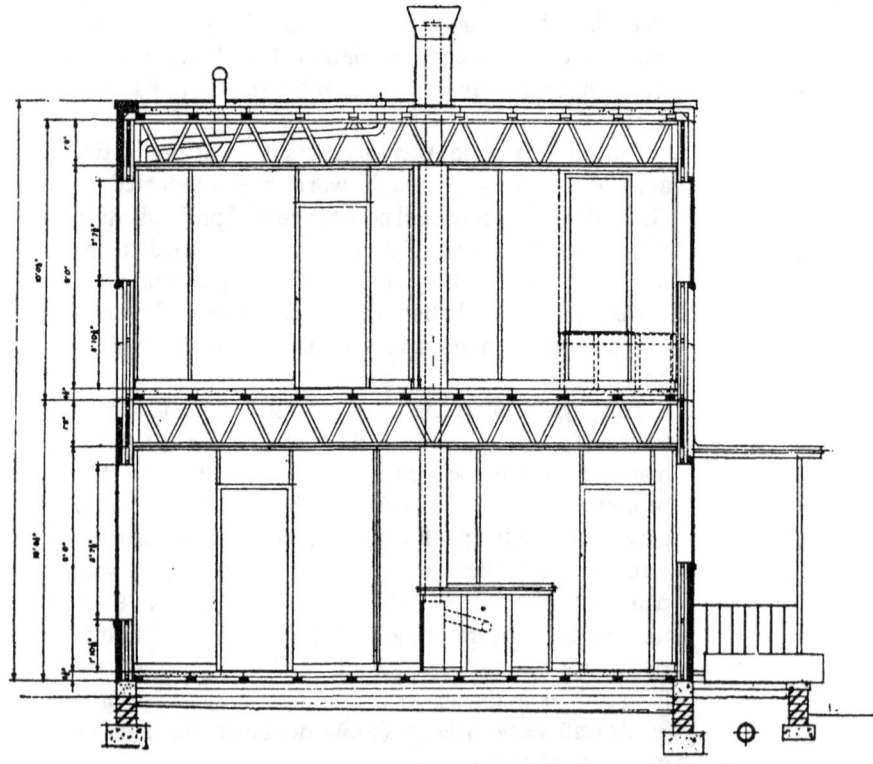

Above and opposite are cross sections of the Keyhouse Unibuilt system prototype houses erected at Coventry. (See plate 33.) Architects: Grey Wornum and Richard Sheppard, FF.R.I.B.A.

The flat roof is constructed of aerated concrete blocks on wood bearers. The floors can be of similar construction or of stressed skin plywood. The 20' floor span (24' for single storey buildings) allows flexibility of planning, as it does not demand load-bearing partitions. The 4' wide external unit, which can accommodate entrant and re-entrant angles, gives flexibility for the general outline of the house.

SKELETON Another British system of steel supporting frame is made by Braithwaite and Company (Engineers) Limited. This is somewhat simpler in form than the Keyhouse Unibuilt, though of heavier steel. The wall chassis are made either 3'2" or 6'4" wide, and are 20' high. The outside vertical members of these are 14 gauge and of a

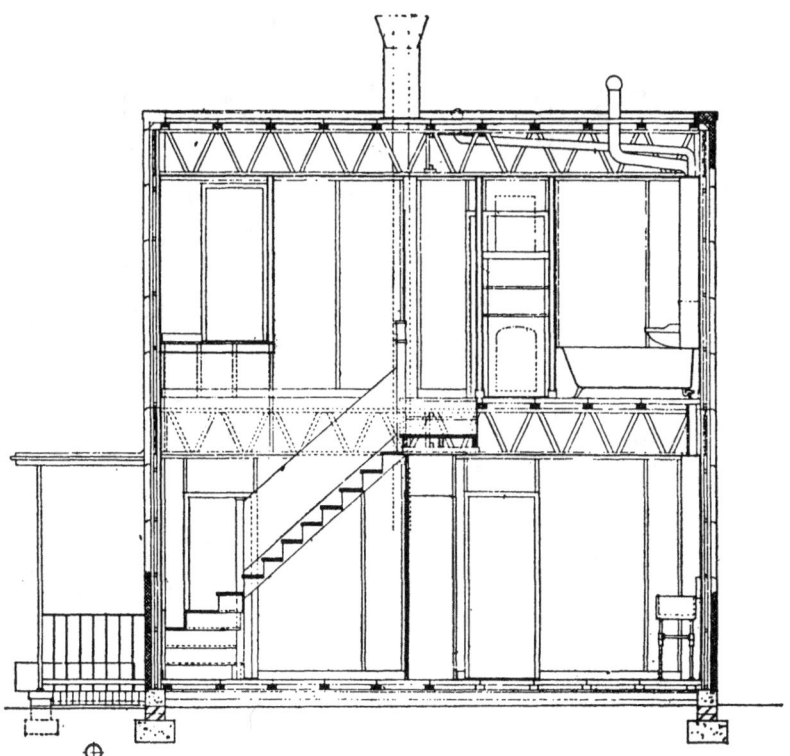

Keyhouse Unibuilt System (see opposite page)

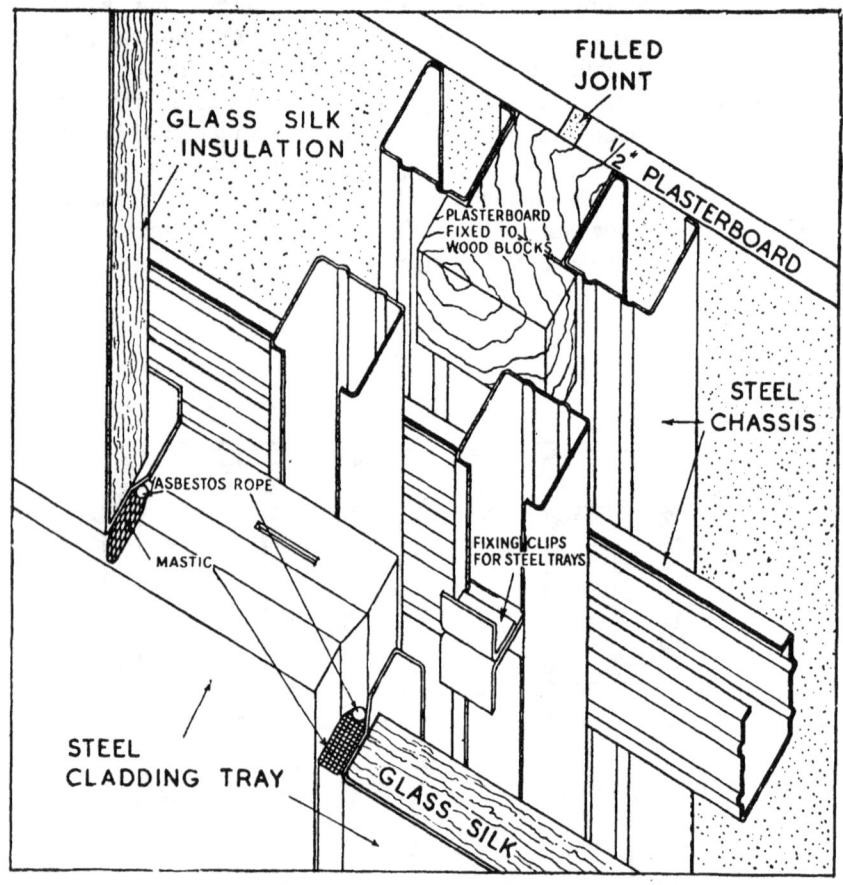

Constructional details showing insulated metal cladding for Second prototype Keyhouse Unibuilt house. Architects: Grey Wornum and Richard Sheppard, FF.R.I.B.A. (See plates 34 and 35)

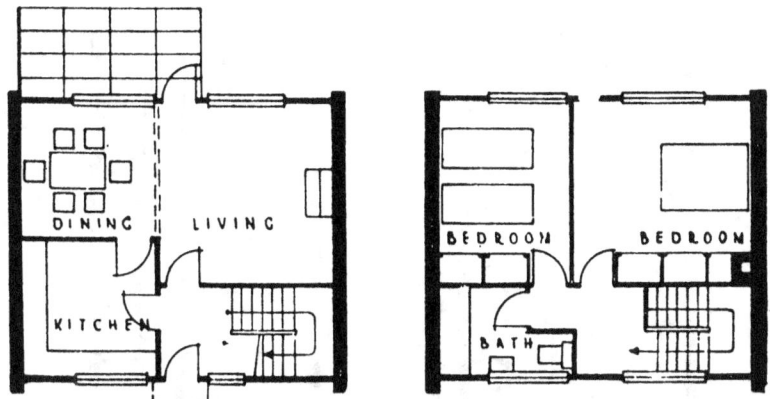

Above, below and next page: Type plans, ground and first floors, of the Braithwaite house. Architect: F. R. S. Yorke, F.R.I.B.A. (See plate 36)
(*This drawing is reproduced by courtesy of Braithwaite & Co., Engineers, Limited*)

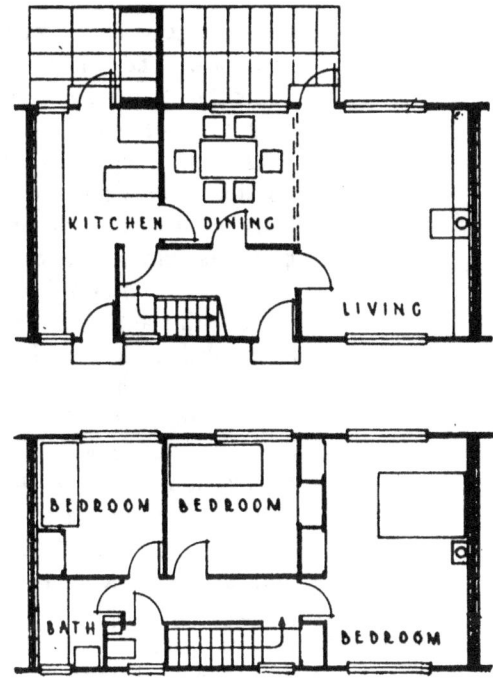

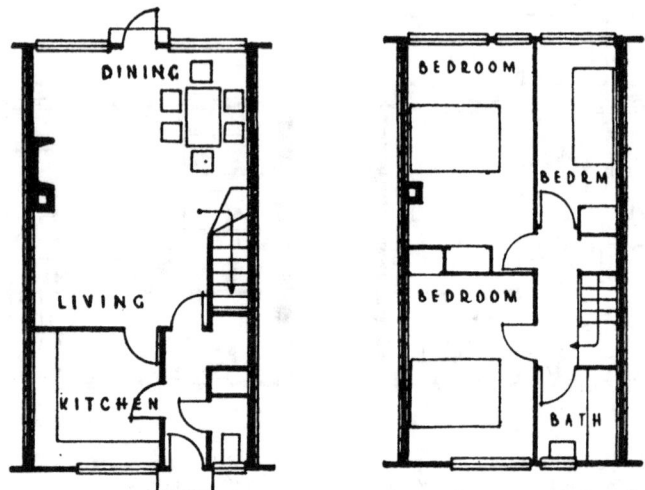

Type plans of the Braithwaite house. (See previous page)

special angle form, making easy combinations of junctions. They are joined by means of bolted angle cleats at 2′ centres. Light horizontal members of 16 gauge steel are spot welded across each chassis at intervals, forming a kind of ladder. They are omitted for door and window openings to be accommodated. Outside sheathing is attached to these horizontal members, either asbestos, aluminium sheet, enamelled steel sheet or even $4\frac{1}{2}''$ brickwork. Wallboard, or plywood with special clip fixings, provides internal lining. Suspended glass wool blanket gives insulation.

Much thought and ingenuity have been given to working out every detail, and the system is extremely elastic for planning. A pair of prototype houses were erected at Hendon in 1944.

Hills Patent Glazing Company Limited, of West Bromwich, have produced two types of steel skeleton, both extremely economical of metal, and permitting a choice of materials for the outside skin. An example **SKELETON** of one system, "Presweld," is the flatted dwellings erected by the Ministry of Works at Northolt, Middlesex. This block of flats in two storeys consists of steel

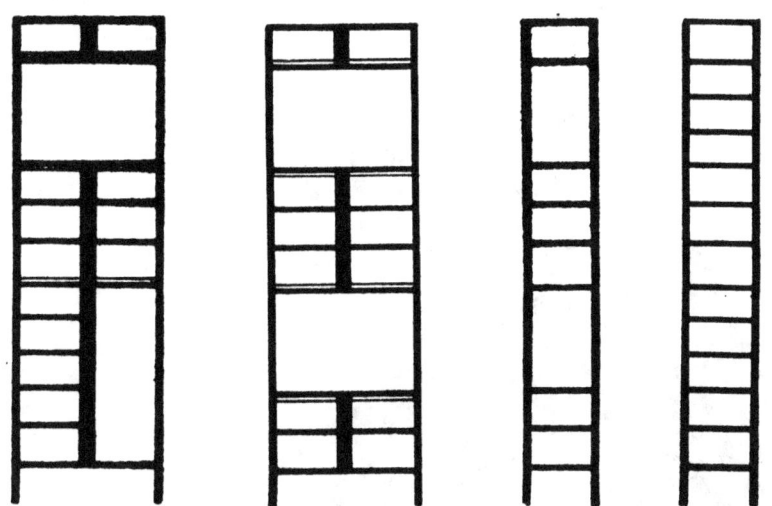

Constructional details of first prototypes of the Braithwaite houses erected at Hendon. Architect : F. R. S. Yorke, F.R.I.B.A. (See plates 37 and 38)

Above : Elevation of chassis types
Below : Plan of chassis assemblies. Axonometric of general system shown on next page

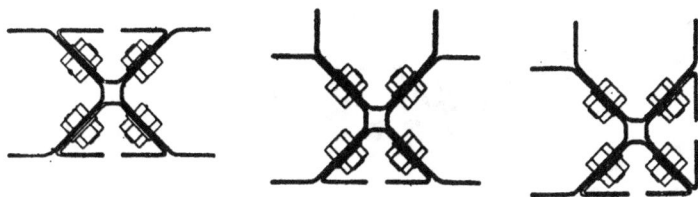

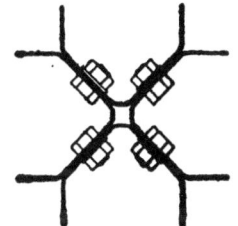

(*This drawing is reproduced by courtesy of Braithwaite & Co., Engineers, Limited*)

Axonometric of general system of prototype Braithwaite houses. (See previous page for plans and elevations)

stanchions, beams, rafters, hips and struts, all made up from combinations of three basic sections of steel, spot welded together. Pre-cast concrete slabs are clipped to the stanchions as outside clothing. Light sheeting rails are run between the stanchions and hold the glass wool insulation and inner wallboard lining. Prefabricated timber floor units are used. The actual time for steel erection of these four dwellings was equivalent to three days' work for six men.

SKELETON   The other Hills Patent Glazing system has been demonstrated at Birmingham. The basic idea was that the lightly constructed skeleton frame could receive a temporary sheathing at first and a facing of brickwork could be erected, later, if desired. The steel construction units are of special design, consisting of a " flat " top and bottom member and a $\frac{1}{4}"$ diameter solid round lacing, spot welded between them. The stanchions are spaced at 3' centres and support the roof trusses above. For temporary outside cladding two thicknesses of asbestos cement sheet are used, secured by bolts to the stanchions. For a permanent skin, $4\frac{1}{2}"$ brick walling is used, tied to the stanchions but separated from them by felt. The inner wall lining and partitions are of coke breeze or foam slag block, plaster skimmed. Roofs are tiled. Like the Braithwaite house, window frames are of pressed steel and cover the full depth of the window openings.

Another alternative in steel skeleton construction is in the use of tubular steel for structural members.
SKELETON   The housing department of the City of Coventry has for some time been experimenting with this, and the
SKELETON   " Arcon " emergency house erected for the Ministry of Works demonstrates its use for roof trusses.

SKELETON   The British Iron and Steel Federation have evolved a system of light construction in steel with Frederick Gibberd, F.R.I.B.A. as architect, and Donovan H. Lee, M.Inst.C.E. as consulting engineer. (Illustrations are shown on pages 45, 46 and 47 and other illustrations of the houses appear on plates 29 to

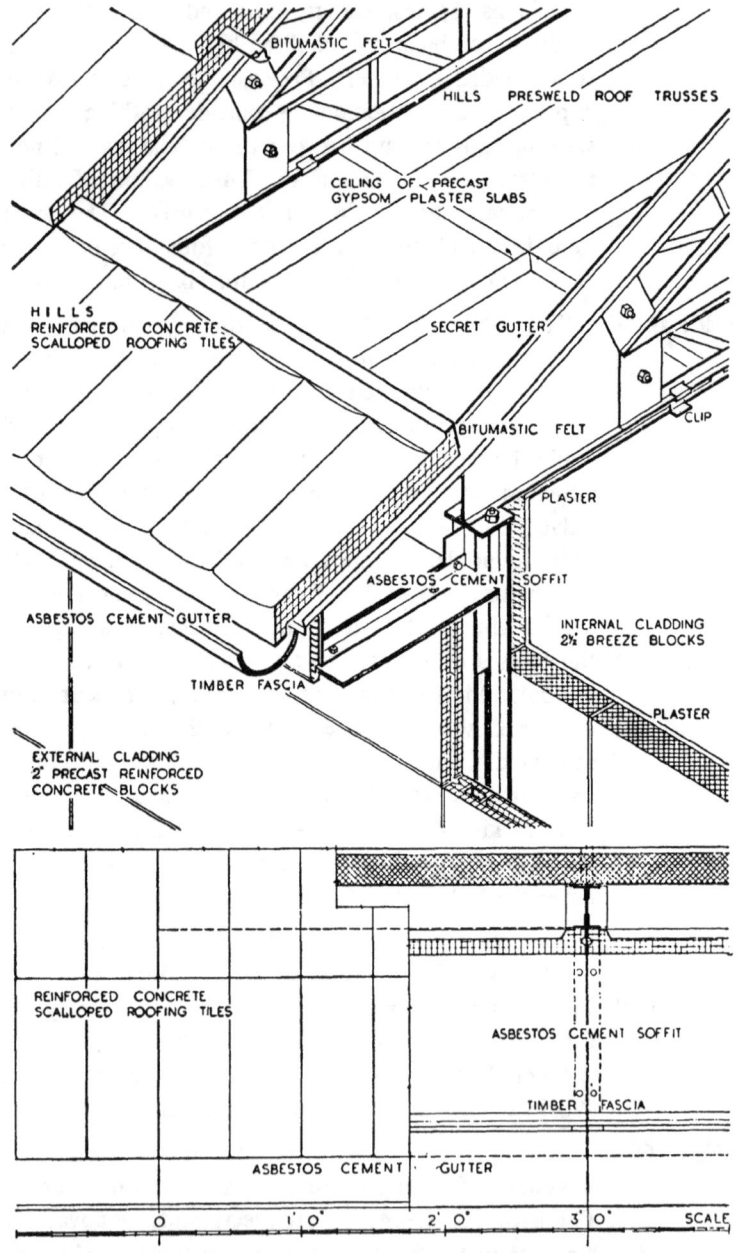

Constructional details of concrete clad houses at Northolt, Middlesex, produced by Hills Patent Glazing Company, Ltd. Roof details shown above. (This diagram and those on the four pages following, are reproduced by courtesy of Hills Patent Glazing Company Ltd.)

WHAT METHODS CAN BE USED? 37

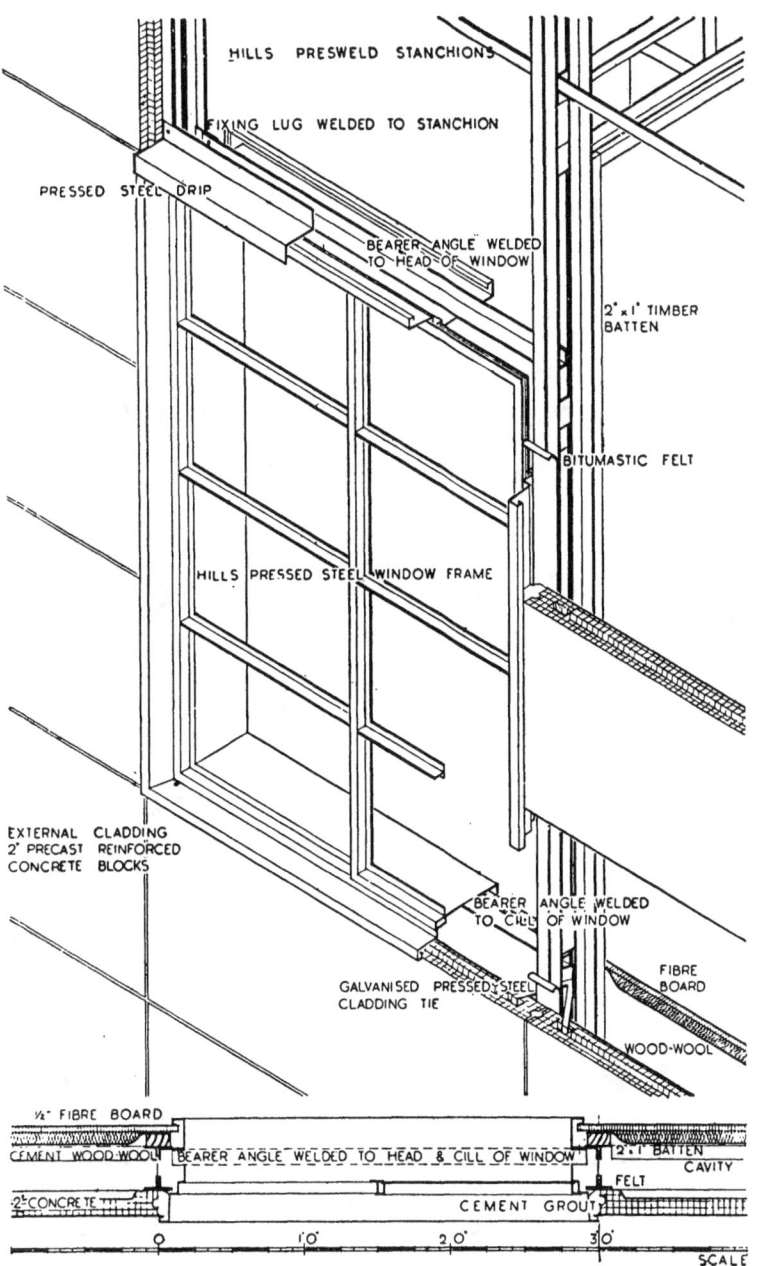

Window details of the Northolt houses

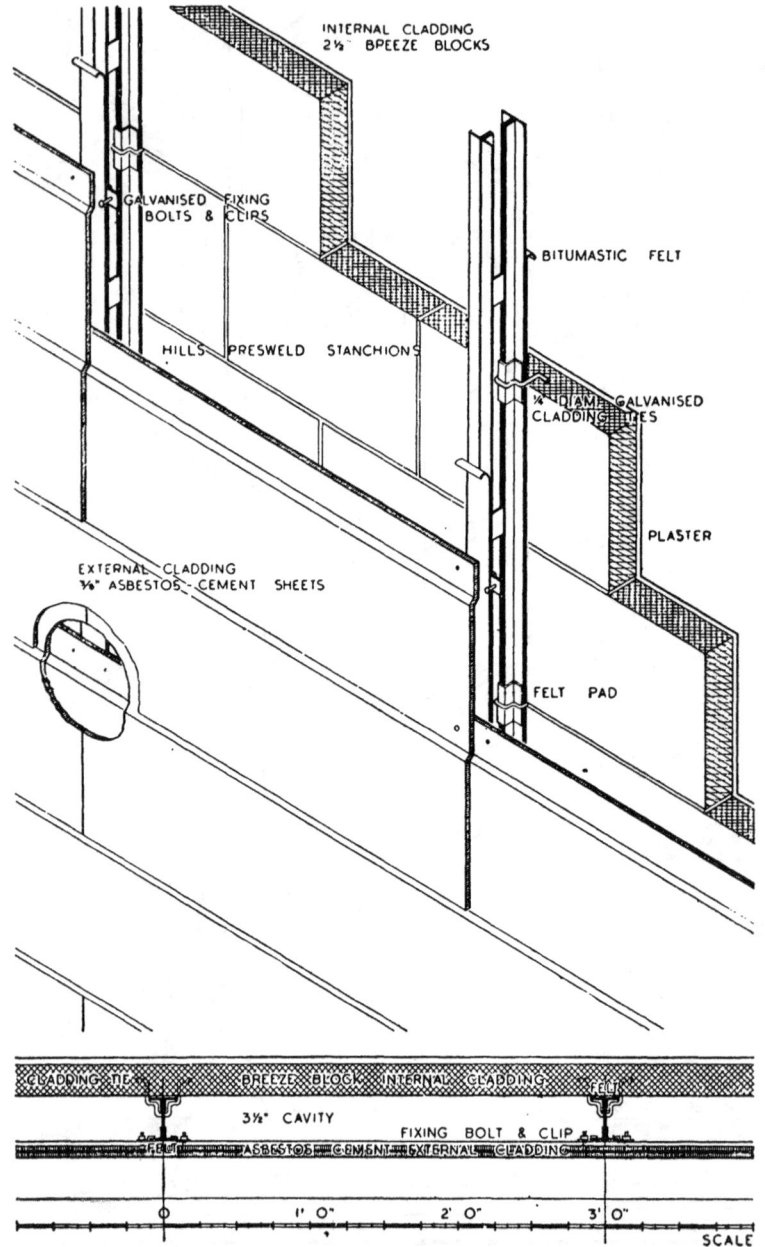

Internal and external cladding of the Northolt houses

WHAT METHODS CAN BE USED? 39

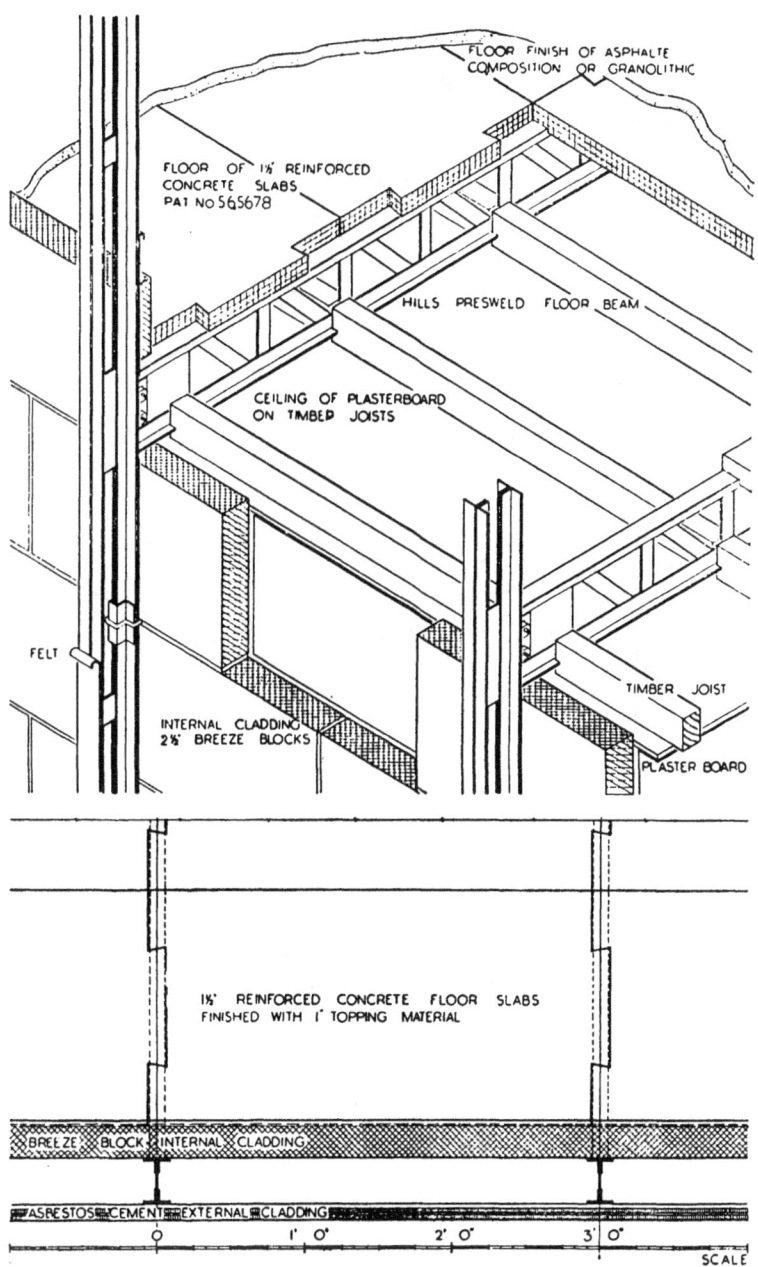

Detail of floors in the Northolt houses

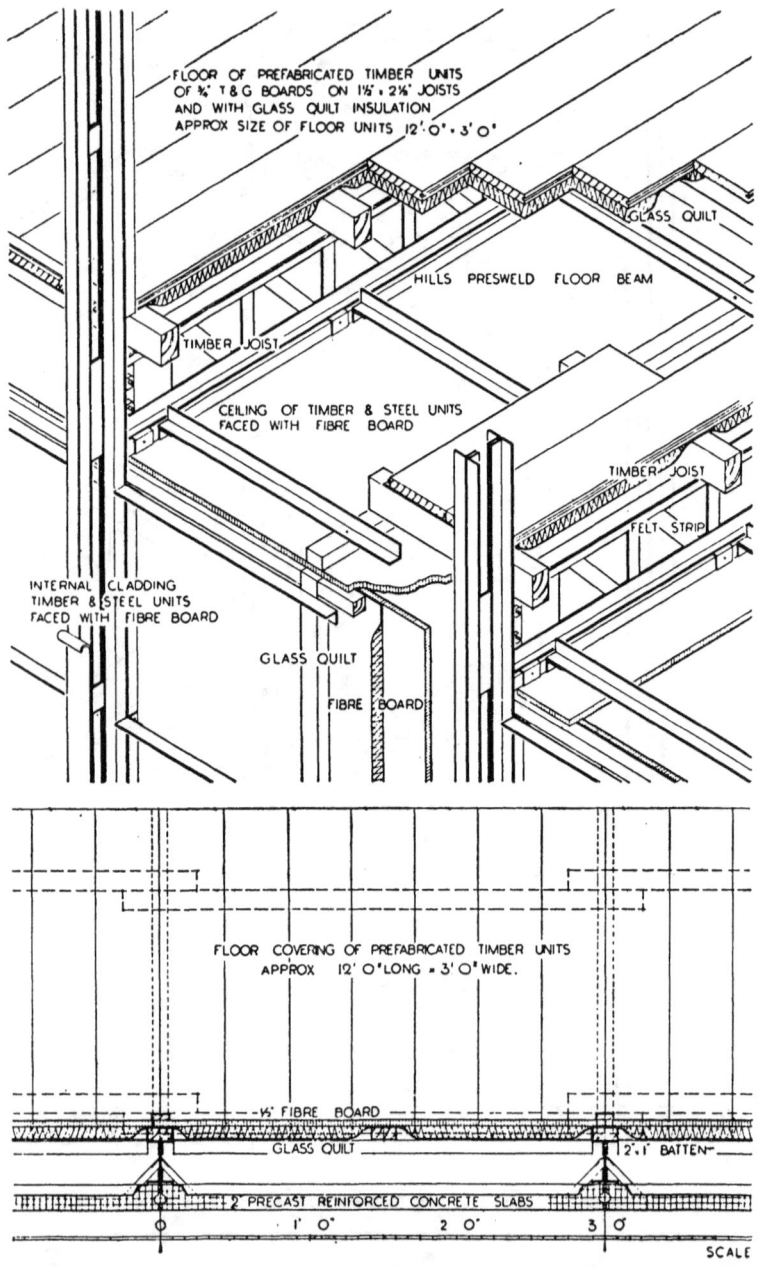

Floor covering of prefabricated timber units in the Northolt houses. (See plate 16)

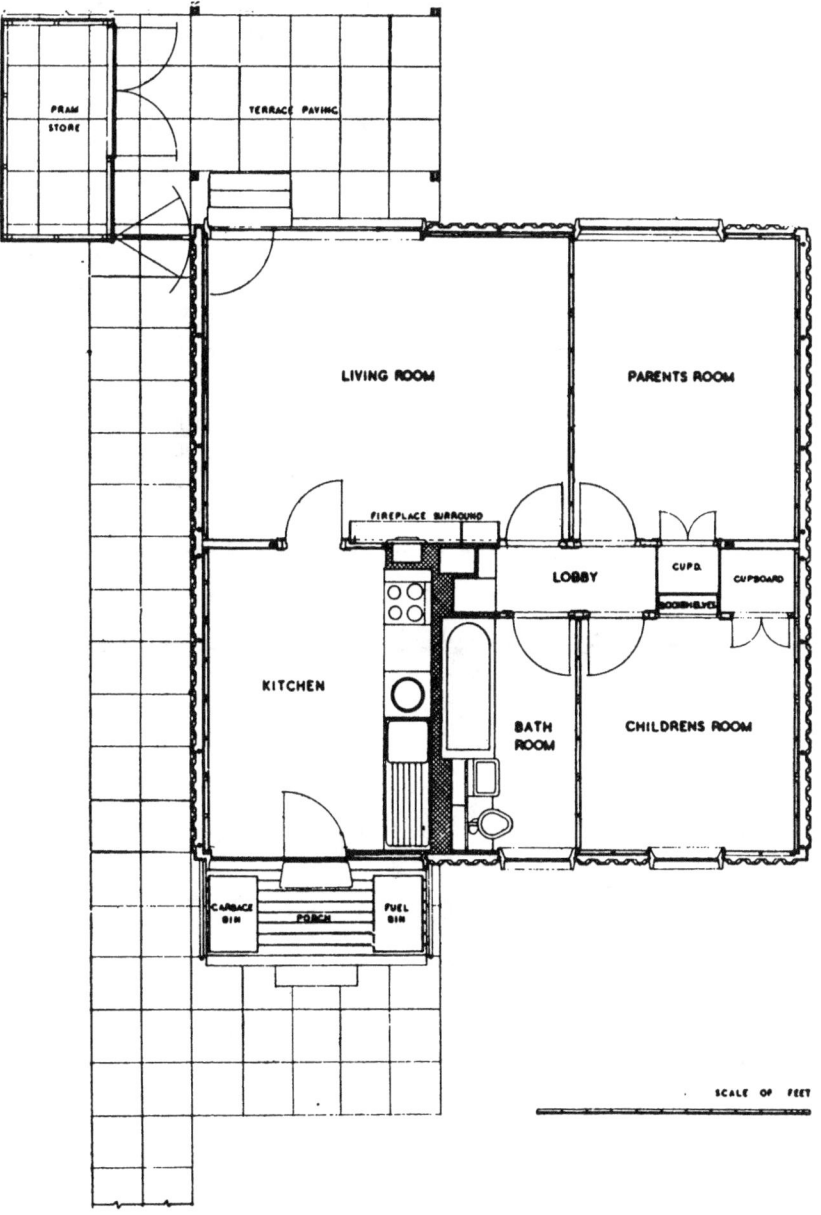

Plan of Mark II demountable house designed by Arcon. (See plate 23)
(*Reproduced by courtesy of Arcon*)

31.) Frederick Gibberd has designed the Howard house (for John Howard & Company, Limited), where the structural elements are of steel, and special tongued wood flooring panels are used. (See illustration on page 48 and also plates 26 and 27.)

CONCRETE

SKELETON

SHELL

Between the two world wars, only four systems of prefabricated concrete houses reached an output of ten thousand dwellings in Britain. They were the Boots pier-and-panel, Duo-slab, Fidler and Winget systems. In the Boots system, pre-cast piers two storeys high were cast on the site and handled by a crane. In the other three systems, pre-cast slabs formed a permanent shuttering to a poured frame or core. They were not in use after the nineteen twenties.

Since those days other systems have been successfully developed, notably the Beaucrete and the Orlit systems.

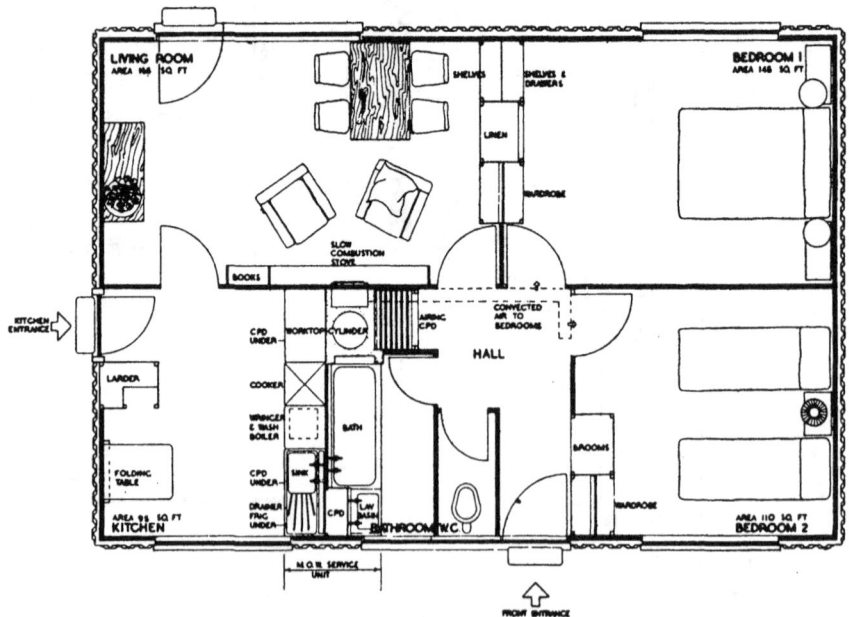

Plan of Mark IV demountable house, designed by Arcon. System of construction appears opposite. (See plate 24)
(Reproduced by courtesy of Arcon)

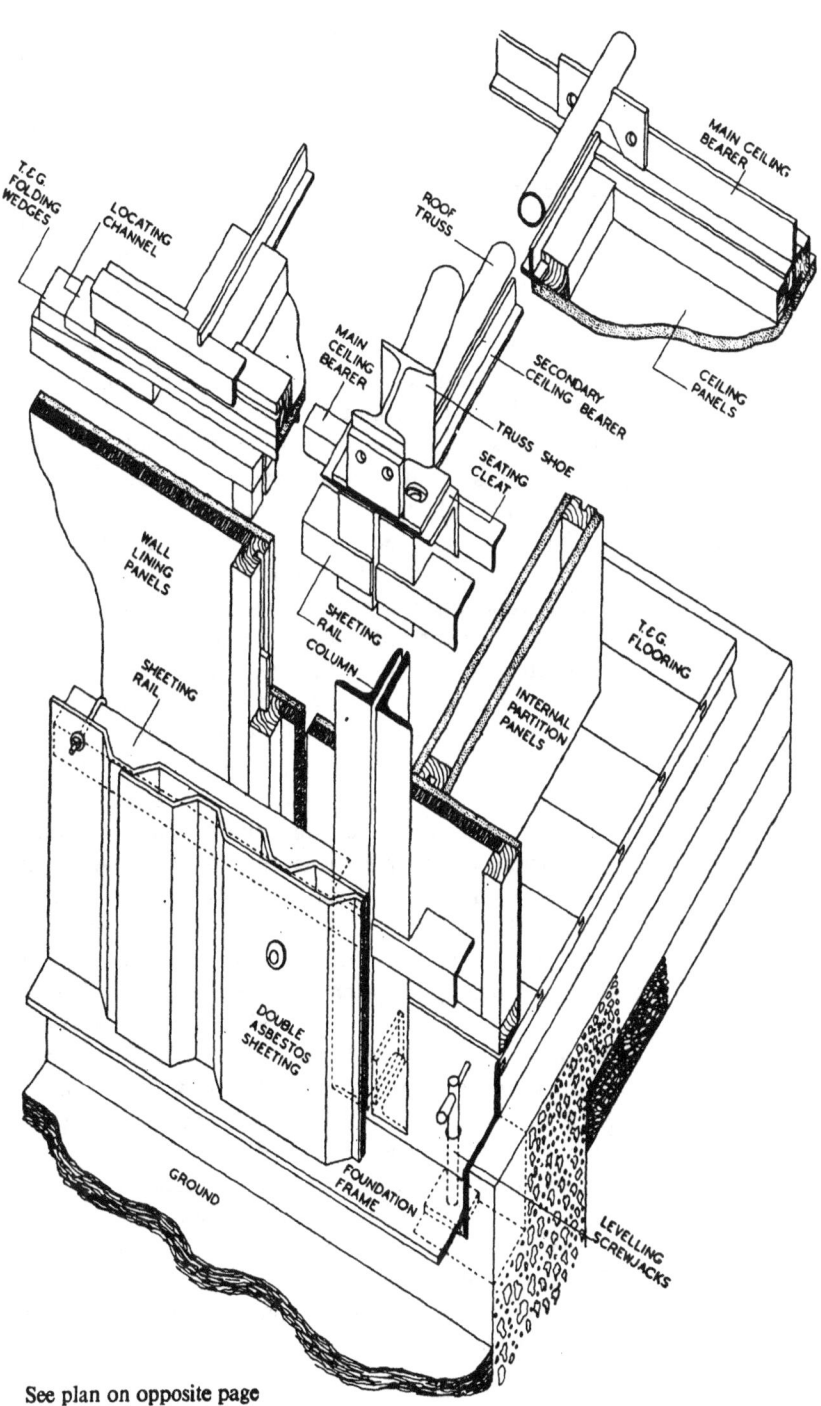

See plan on opposite page

SKELETON	Both employ a pre-cast concrete frame and a double skin panel facing, kept in place by metal ties bridging across the two panel walls.
SKELETON	The Boots system has double vertical frames, one behind the other, with bolted junctions top, middle and centre. Either timber or slab concrete floors may be used. The outer covering is formed by thin horizontal concrete slabs, fixed dry between joints of felt. The inner skin of the wall is of foam slag block, plastered *in situ*. The concrete construction is on the heavy side.
SKELETON	The Orlit system has a 2½" thick concrete masonry block on the outside with a stucco face cast integrally, and a 2½" foam slag inner lining, plastered. A pair of these houses has been erected at Colnbrook. In both systems pre-cast concrete beams connect with the vertical supports and hold up the floor.
SKELETON	The system evolved by Sir William Airey at Leeds consists of pre-cast vertical concrete supports, into which wood fillets are inserted on the inside face as fixing grounds for the wallboard that lines the interior. Concrete slabs are fixed on the exterior face by means of wire attachments taken round these supports.

In all three systems added rigidity is given to the concrete frames by the concrete panel filling.

In comparing early pre-cast concrete systems in different countries we find in Britain a preponderance of concrete masonry systems and pier and panel; in Germany, an almost exclusive use of load-bearing units, either in block form or " Massiv " slab form. America displays a much more adventurous use of the material, the most successful system commercially being that of simple channel-shaped storey height units. American development of this material began in the first decade of the century, with Grosvenor Attebury's experiments. Such pioneer systems attempted to fabricate the material in huge and complex pre-cast slabs, a tendency that has persisted throughout the development of concrete units. In the nineteen twenties load-bearing panel systems began to appear, mostly of tray or pan shape. This development continued up to the outbreak of the second world war. Units of this type were usually of storey height and from one to four feet wide. They

WHAT METHODS CAN BE USED? 45

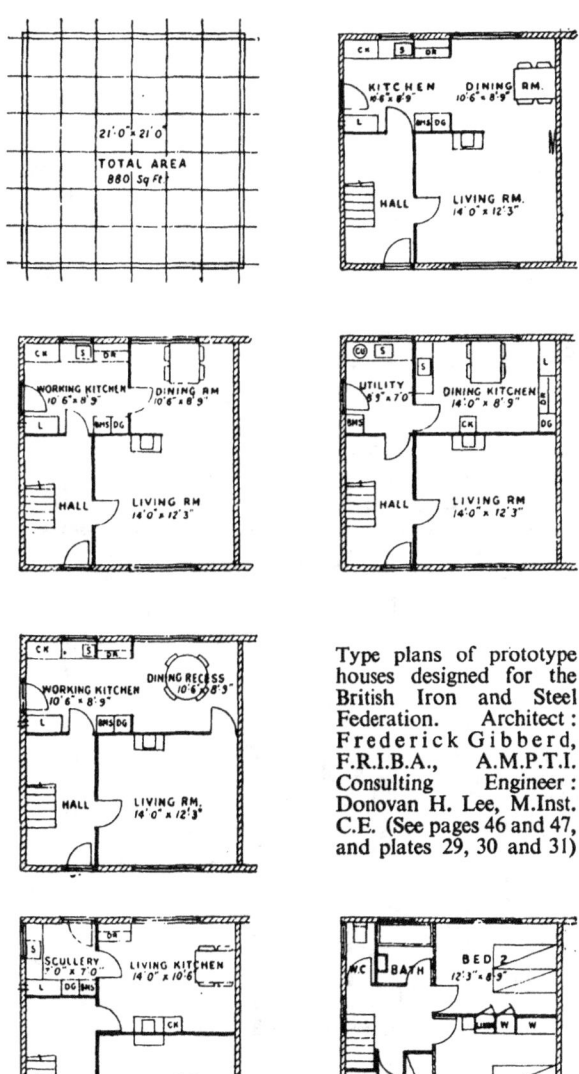

Type plans of prototype houses designed for the British Iron and Steel Federation. Architect: Frederick Gibberd, F.R.I.B.A., A.M.P.T.I. Consulting Engineer: Donovan H. Lee, M.Inst. C.E. (See pages 46 and 47, and plates 29, 30 and 31)

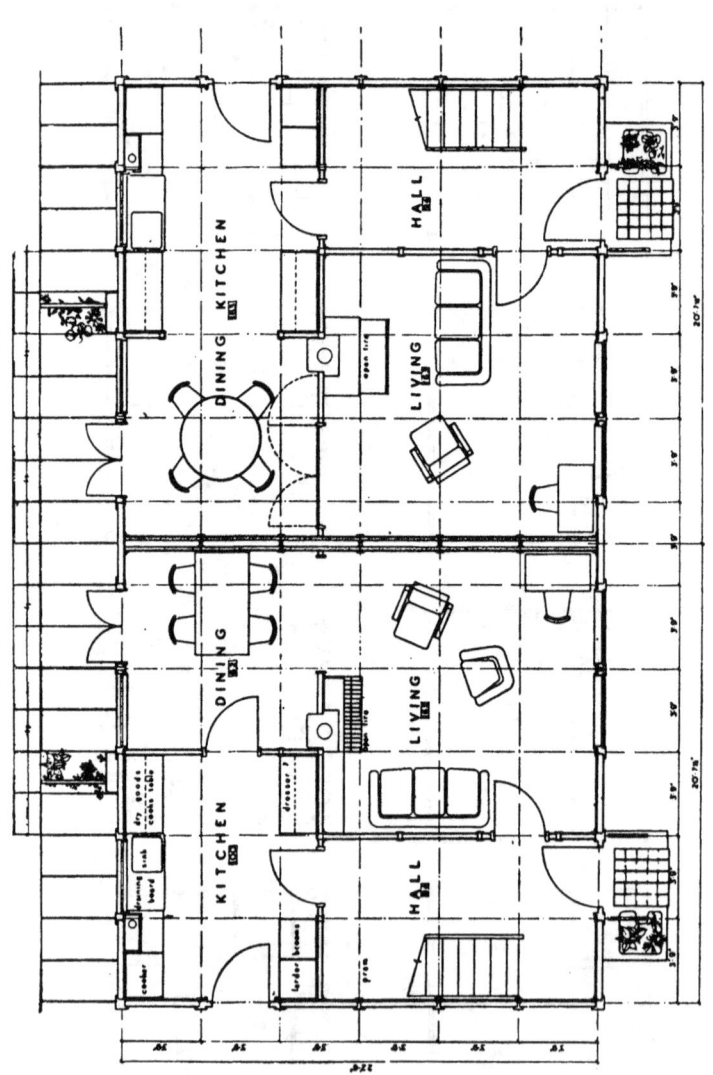

WHAT METHODS CAN BE USED?

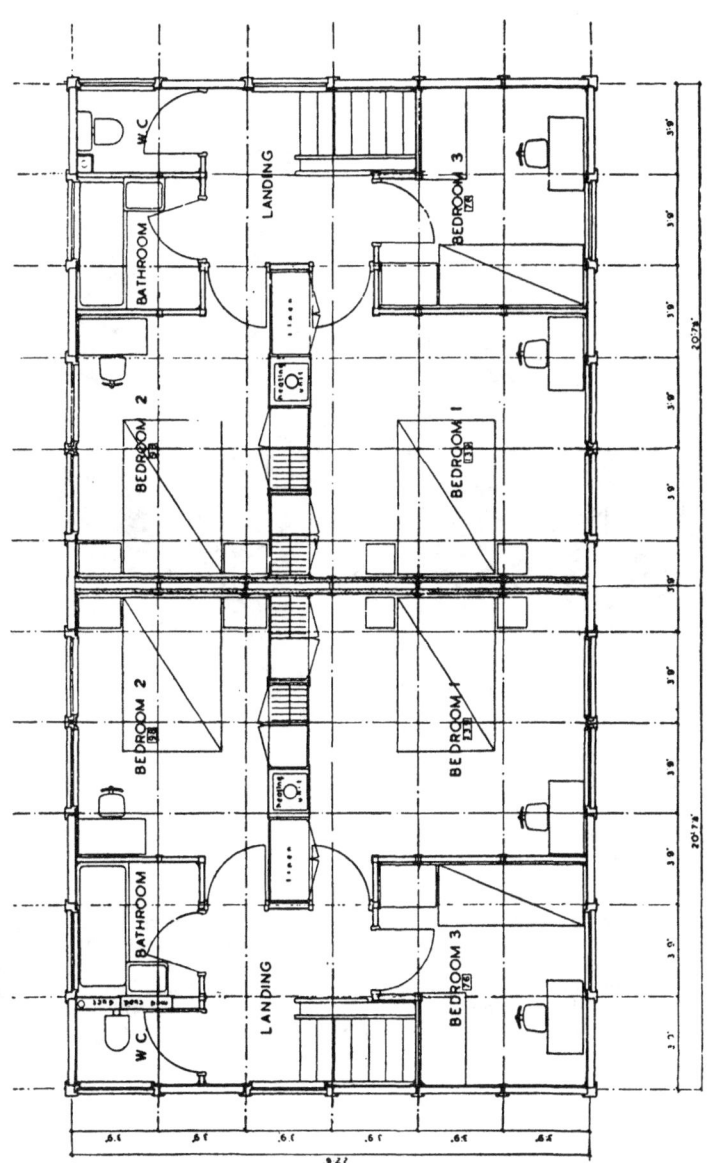

Ground and first floor plans of a pair of prototype houses produced by the British Iron and Steel Federation appear here and opposite. Architect: Frederick Gibberd, F.R.I.B.A., A.M.P.T.I. Consulting Engineer: Donovan H. Lee, M.Inst.C.E. (Ground floor plan opposite, first floor plan immediately above) (See plates 29, 30 and 31)

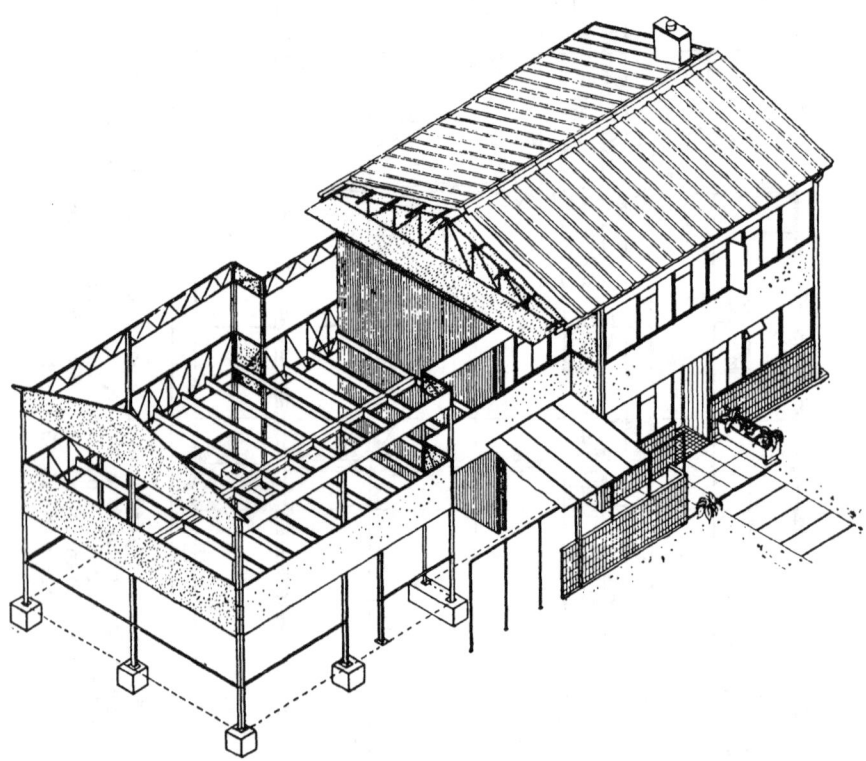

Diagram of a pair of Type A Howard houses erected at Datchet, Berkshire, showing main structural elements. This house is of permanent and completely " dry " construction, based on a 10' grid with self-contained kitchen and bathroom units comprised in the central recessed portion. The structural elements are of steel ; specially tongued wood flooring panels are used and a considerable amount of insulation is provided to walls and roof.
Architect : Frederick Gibberd, F.R.I.B.A., A.M.P.T.I. (See plates 26, 27 and 28)
(*Reproduced by courtesy of John Howard & Company, Ltd.*)

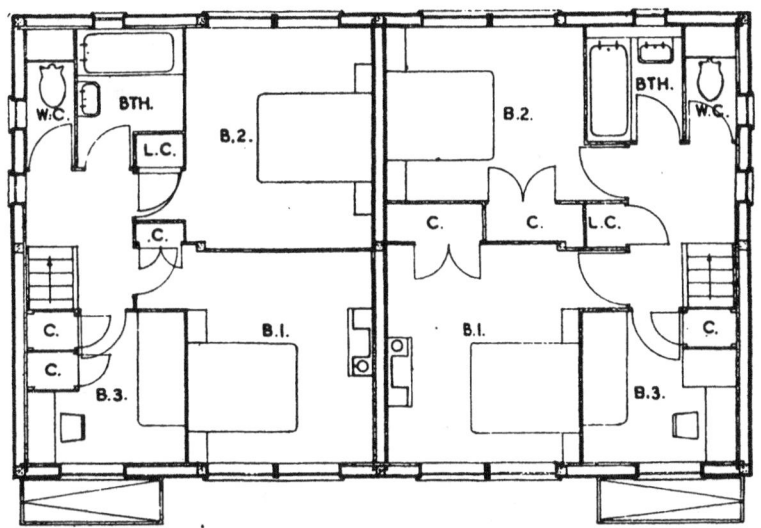

FIRST FLOOR PLAN.

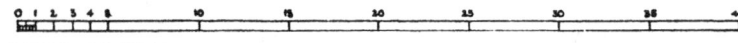

SCALE OF FEET.

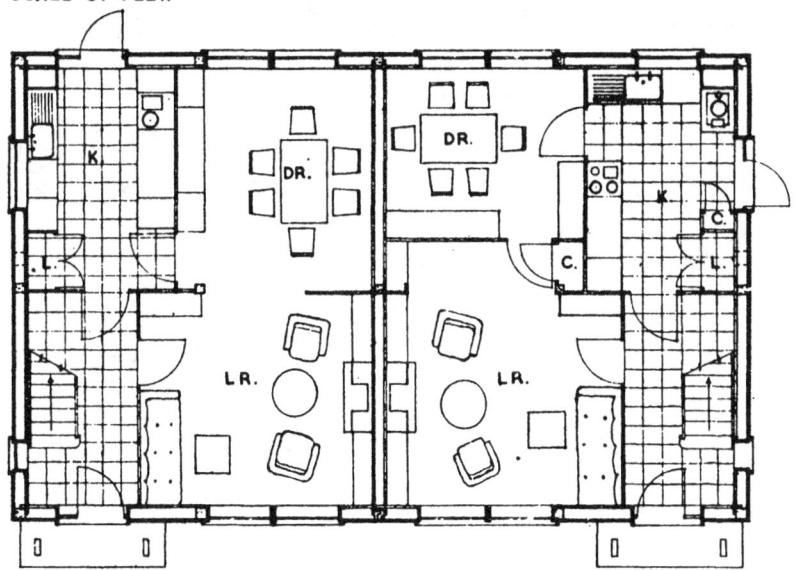

GROUND FLOOR PLAN.

Ground and first floor plan of the Type 5/1 (all-electric) and Type 5/2 (gas and coke) Orlit concrete houses built at Colnbrook, Middlesex. Architect: E. Katona. (See page 44 and plate 15)
(*Reproduced by courtesy of Orlit Limited*)

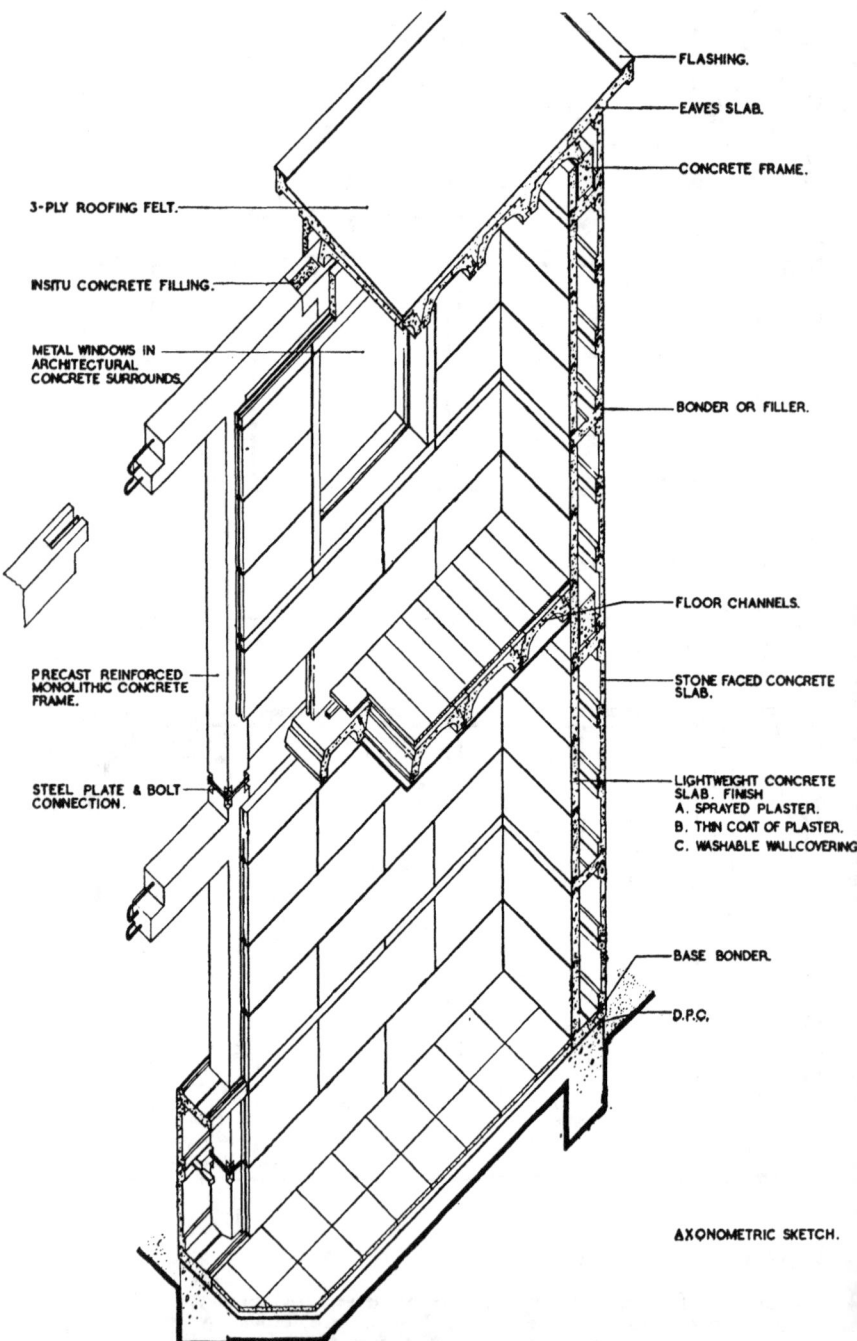

Construction details of concrete houses built at Colnbrook, Middlesex, by Orlit Limited.
Architect: E. Katona. (See pages 44 and 49, and plate 15)
(*Reproduced by courtesy of Orlit Limited*)

were more readily handled than the monolithic structures, but the weight of the units always involved mechanical equipment and so limited the range of handling. Panel and stud systems are plentiful, but were not very successful after 1930.

In Germany, extensive experiments in concrete house construction took place between the wars, parallel with and sometimes in conjunction with the use of steel. Pumice, clinker, foamed slag and various other aggregates were used to produce lightweight concrete. Penetration by rain was not uncommon, but many successes were achieved. While blocks of the more usual size were used, some large scale slab construction was tried out, of storey height and 4" thick, for partitions. Such heavy units entailed special plant and organisation, which limited their use. Development took place in prefabricated insulation slabs, sometimes reinforced with stiffeners. These slabs ranged up to 3" thick and in size were about 4' × 8'. The materials that were used had been known in Britain for many years, but our own structural methods had not evolved to the stage when we could use them. One of the most interesting of Germany's experiments was in prefabricated flat construction. An experimental housing scheme at Munich was carried out in many types of construction on the basis of a standard plan, and comparative costs were obtained.

Numerous types of concrete block construction have since been evolved as an improvement on brick construction. It has not yet been scientifically determined whether lightweight units of larger size could be more rapidly laid than bricks or whether larger and heavier units made with tackle or machinery would be more efficient. Certain types of hollow, inter-locking shapes provide insulation and keep out weather, but are handicapped by lacking an attractive and weather-resisting face. Large scale wall sections have been cast complete on tables or flat on the site. Door and window surrounds, ducts for services and air spaces for insulation have been provided. The range of machinery and tackle for handling such large sections is considerable. Pre-cast frame work in concrete has been mainly used where timber or steel were in short supply. The provision of foundations in pre-cast elements, piers and beams, is common practice for demountable timber defence housing in the U.S.A. The use of concrete beams and hollow units with poured filling for floors in conjunction with normal masonry walling is common practice everywhere. In Britain it has received an impetus from wartime shortage of shuttering material, which has precluded the forming of slabs. While the *in situ* factory can deal with the heavy forms of prefabricated concrete, there is likely to be considerable use of the reinforced lightweight concrete blocks in " composite " construction, especially for floors and

SKELETON roof. Such a technique was most successfully produced by Mopin in France before the war. He used a light steel skeleton structure, with an outer cladding of large, thin, vibrated concrete slabs, faced in their moulds with small pebble or marble chips. Such construction was applied to multi-storey blocks of flats at Drancy in France, and was later employed for a large scheme of flats at Leeds. The reconstruction as originally designed by Mopin did not comply with the building regulations current in England in the early nineteen thirties for multi-storey building, and so lost much of its prefabrication value.

Concrete as a structural skin is too heavy a material for either handling or transport to have much future in prefabrication. But there is an American system with a special method of reinforcing, which allows the whole side of a single storey house to be cast at one

SHELL time. This complete wall is $2\frac{1}{2}''$ thick and includes door and window openings. It is lifted without cracking or damage, by means of cranes with vacuum attachment, and lowered on to a lorry for transport to the site. Such houses have been built for the U.S.A. Maritime Commission's Housing at Tampa, Florida ; they are $25' \times 30'$ and include two bedrooms. The special method of reinforcement is by pre-stressing the reinforcing bars. These bars are coated with a thermoplastic material, and four days after pouring the concrete, heat is applied to them by electrical power, which melts the plastic coat, thus releasing the bonding of the bars to the concrete. This allows the stressing of the bars by means of nuts on the threaded ends, the whole operation to each bar taking only 30 seconds. The final compressed state of the concrete due to this prestressing prevents any cracking. (See plate 8).

SHELL Another example of concrete structural skin is afforded by a Swedish system. Lightweight vertical reinforced slabs about 9' high and 18" wide are set up on end, and through the holes running horizontally across them at 2' centres four steel wires are threaded on a kind of bodkin. These wires are then stressed with a rachet, thus pressing all the slabs tighter together. Cement is then inserted under pressure, through special apertures, into the holes containing the wires, thus setting them permanently under stress. It is claimed that this system avoids all likelihood of the concrete cracking or the joints of the slabs opening through expansion. The floors and roofs are constructed in a similar manner and the upper floor walling units are placed over the lower. The external face is then treated with paint or waterproof distemper. Such a system is known as *post-stressing*, and, to make such construction practicable, a lightweight aggregate or aerated concrete is needed. Concrete of such character, apart from saving weight, can give good thermal insulation.

Experiments are being conducted in Britain with lightweight aggregates, such as foam-slag, particularly for poured *in situ* methods, and also for processes of foaming or aeration.

SHELL Successful experiments have been made at Glasgow, by Mr. J. H. Ferrie, L.R.I.B.A., the chief housing architect of the city, in the pre-casting of single reinforced foam-slag units, $10' \times 8'8\frac{1}{2}''$ and 6" thick.[1] The weight of such concrete can be varied according to the strength required; for practical purposes the weight may be considered about half that of ordinary concrete. But a structural skin can be provided by materials other than concrete.

---

[1] "Experimental Flatted Houses," designed by J. H. Ferrie, and W. Kerr, *Architects' Journal*, August 17, 1944.

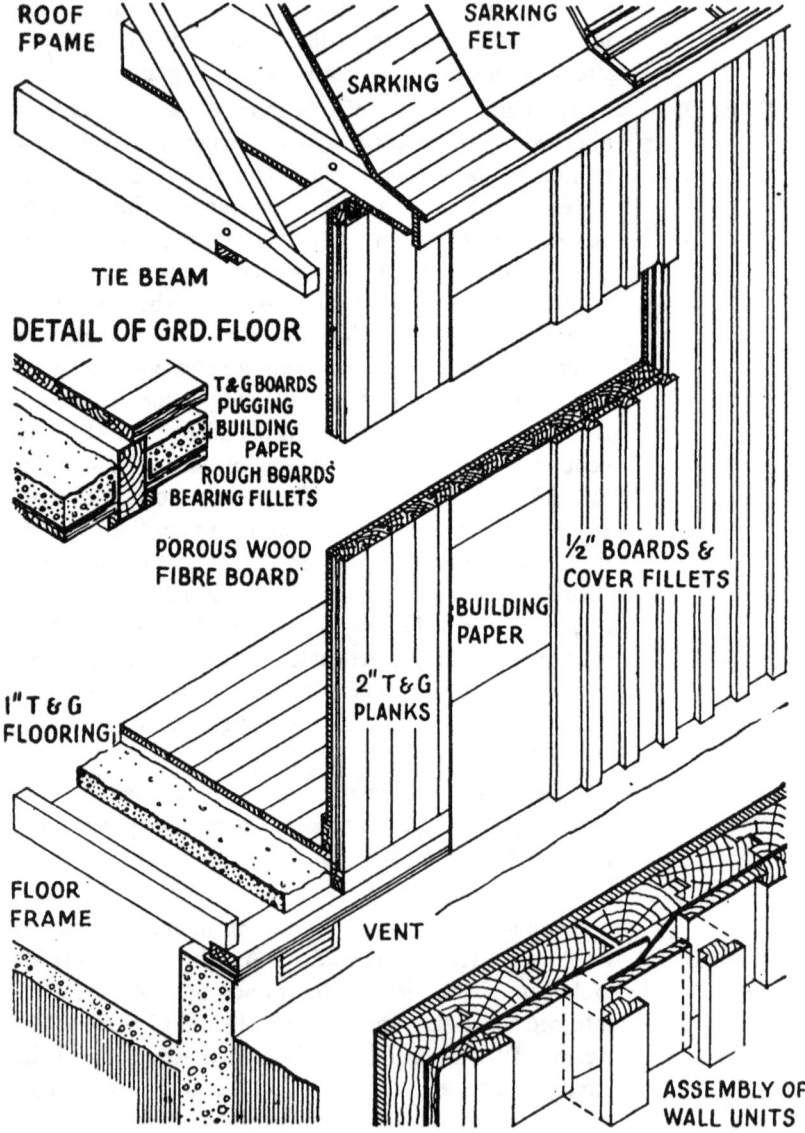

Constructional details of Scano I house, produced by Scanhouse Ltd. Drawing by George Fairweather, F.R.I.B.A.
(*Reproduced by courtesy of the* Architect and Building News)

TIMBER

Systems of the structural skin type are well served by timber, for it may be used either in the form of solid boarding, framed panels, or stressed skin plywood. Various solid boarding systems were in use before 1939. For example, five hundred timber houses built in Dundee, Scotland, in 1937, by Tarran Industries, used 3" tongued and grooved solid vertical boarding, with horizontal cedar weather-boarding (or siding) on the outside, a $\frac{3}{4}$" cavity, and wallboard lining on battens for the interior. A Swedish solid boarding system, the Scano, uses less timber, and has a core of 2" thick tongued and grooved vertical planks covered outside with waterproofed paper and then faced with $\frac{3}{4}$" vertical boards. Narrow wood fillets are sometimes used to cover the joints of these boards. Fibre board is used on the inside as wall lining.

A system commonly used in Sweden, and which is economical of timber, is the framed panel type. Such Swedish systems as the Ibo use substantial framed panels faced on both sides with 1" thick boarding and filled inside with rammed shavings. Boarding and cover fillets are used for an outer skin, or plaster rendering on lathing. Framed panel construction has also been extensively used in the United States, particularly by such manufacturers as Gunnison and the Douglas Fir Plywood Association. America presents a greater variety of timber building techniques than any other country, and has developed more experiments in prefabrication. But although the making of " frame " timber houses is a traditional practice in the United States, the completely portable house is not so old. In the past, the necessity for it seldom arose, because of the abundance of timber in North America : there were nearly always local supplies to draw from for building. Early examples of complete portable timber houses date from the eighteen nineties. Hodgson, 1892, and Alladin Bay, 1906, were typical prefabricators. The standard of construction was much better than comparable examples of portable timber buildings in this country. In general, it was customary for wood houses to be "sectional," and "pre-cut." Such firms as the E. F. Hodgson Co., (Sectional), and Sears Roebuck (Pre-cut) House, each sold over one hundred thousand dwellings during the last forty years.

The relatively familiar systems of solid boarding or framed panels have been widely used in timber producing countries; but an entirely new system of construction has been invented in connection with plywood. Known as "stressed skin" construction, it was developed by the Forest Products Laboratory at Madison, Wisconsin, U.S.A. The basic principle is that plywood, glued both sides of a rib core as a skin, assists to carry the loads imposed on the walls. Under wind pressure the outer skin tends to compress and the inner skin to stretch. The plywood resists these stresses as well as acting as a pillar to support the floors and roofs. These plywood faced panels are generally 8′ high by 4′ wide, and the outer plywood is resin-bonded to resist the weather. The whole panel is two or more inches thick. The gluing of the plywood to the core of small wooden ribs makes the panel completely rigid and under stress it acts as one unit. Insulating material can be placed between the two skins, and floor and roof panels can be similarly constructed. Many houses have been built on these principles, particularly in Los Angeles, California, and "stressed skin" has now become an established technique. (It has been used considerably in aeroplane construction, especially for the "Mosquito".) The first example of stressed skin plywood construction for house building in this country is the Jicwood bungalow at Weybridge, Surrey, designed by Richard Sheppard, F.R.I.B.A. It was built to satisfy the floor area requirements of the Mininstry of Works, and is in eight sections. Walls, floors and roof are all constructed from material $1\frac{3}{8}''$ thick, which is made up of $\frac{1}{8}''$ sheets of plywood, with expanded rubber plastic between: a wood-faced sandwich with an insulating core. Other materials are available as alternatives for forming this core. The insulating properties of such walling are equivalent to those of an 11″ cavity brick wall. It is estimated that 200 man hours are needed

SHELL

WHAT METHODS CAN BE USED? 57

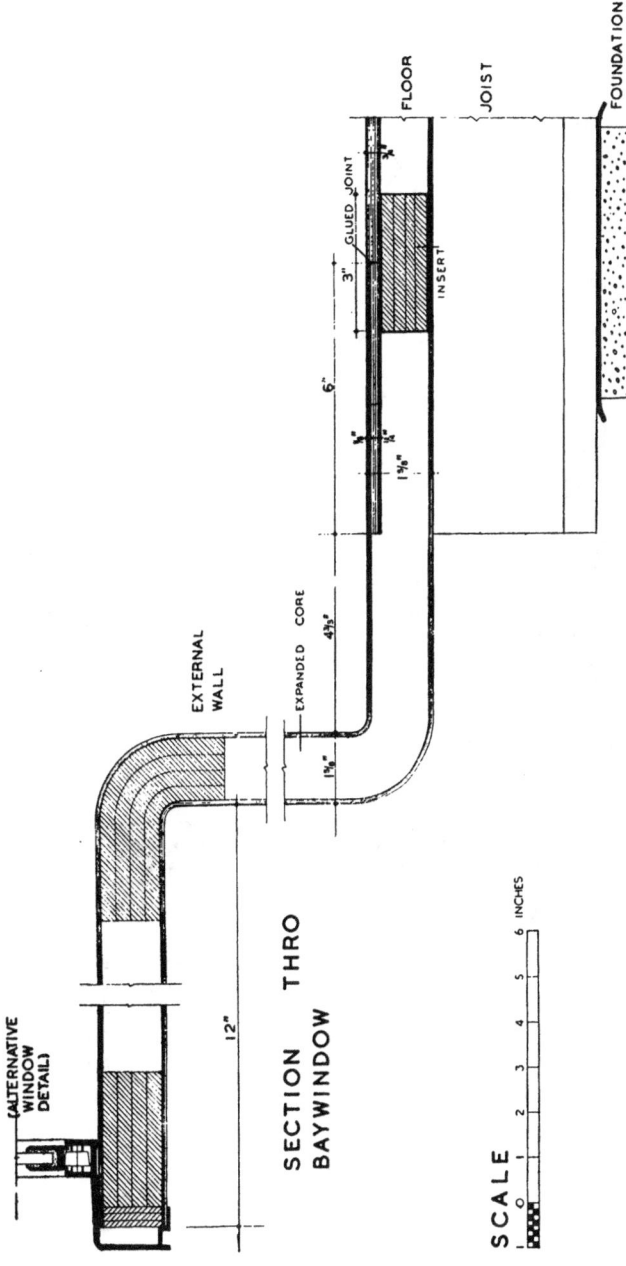

Construction details of Jicwood prototype house. (See pages 58 and 59)

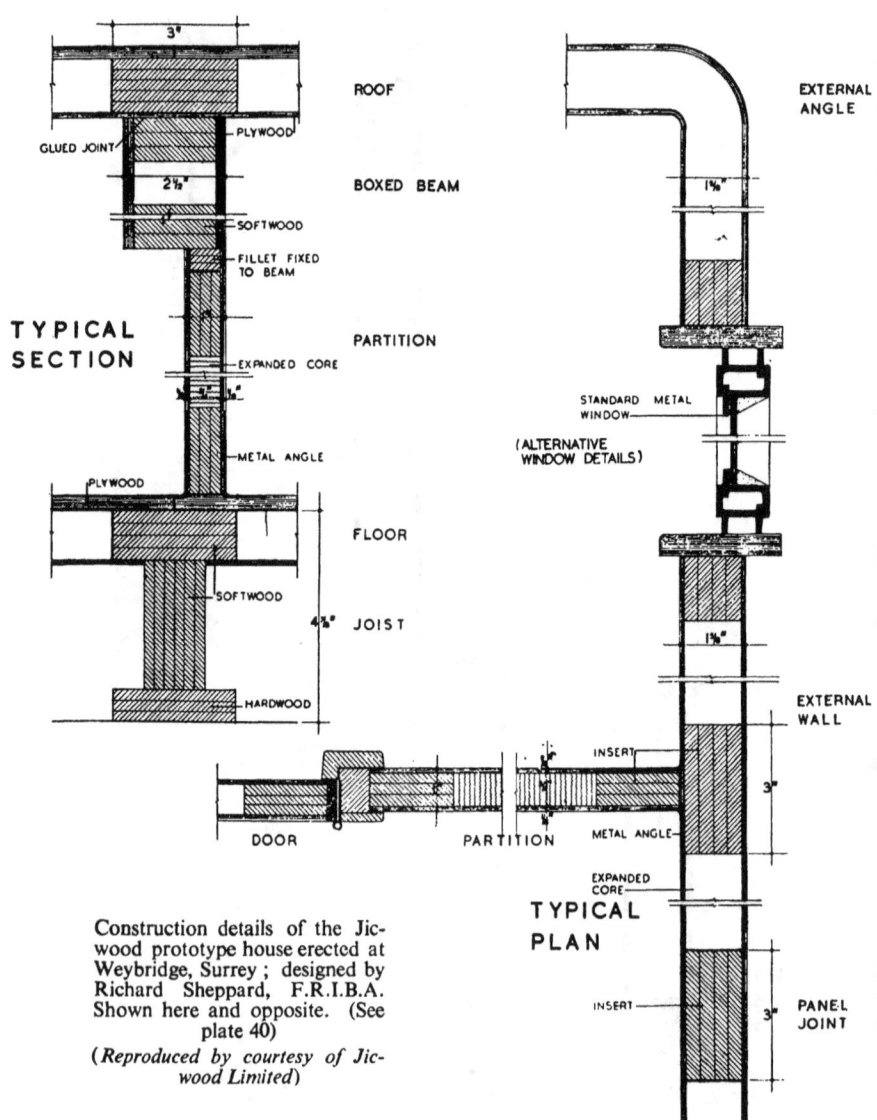

Construction details of the Jicwood prototype house erected at Weybridge, Surrey; designed by Richard Sheppard, F.R.I.B.A. Shown here and opposite. (See plate 40)

(*Reproduced by courtesy of Jicwood Limited*)

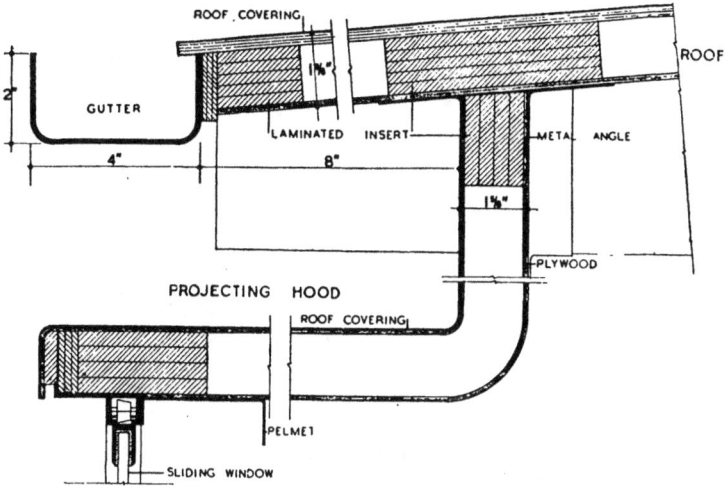

Construction details of the Jicwood prototype house. (See opposite and page 57)

for erecting the Jicwood house, and when demounted a loss of only 5% of material occurs on re-assembly. For its construction panels of a maximum size of 8'×4' are joined in the factory to form larger sections, such as complete walls. Apart from these sections, the roof is supported by timber boxed beams at 4' centres, and the floor by timber joists at similar intervals. The internal partitions are made up in lengths and assembled on the site by gluing. All joints are formed by a spline inserted between panels. Angles between walls and floors are formed by a one-foot section shaped in the press. The whole structure can be transported to the site by lorry.

Mobility is essential for housing schemes that have to be built at speed. The use of timber construction for the great output of houses in the U.S.A. has been due to the fact that the material is abundant and available on the spot, steel being withheld for war purposes and concrete failing to meet the requirements of mobility. In the Tennessee Valley scheme the need was for temporary workers' houses to be speedily

60 HOUSE OUT OF FACTORY

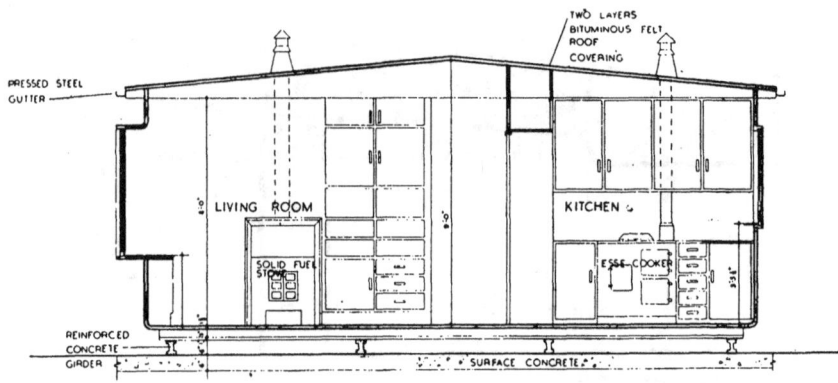

Section and plan of Jicwood prototype house erected at Weybridge, Surrey; designed by Richard Sheppard, F.R.I.B.A. (See p'ate 40)
(Reproduced by courtesy of Jicwood Limited)

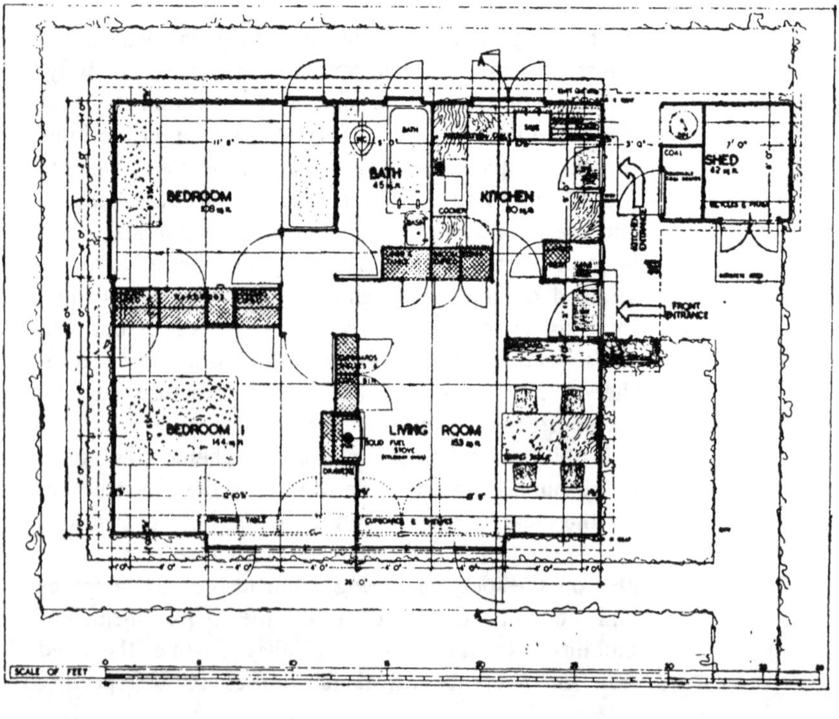

WHAT METHODS CAN BE USED ? 61

Tennessee Valley Authority demountable multi-unit houses elevation. (See plates 9, 10 and 11). Plans appear on next page
(*Reproduced by courtesy of the* Architectural Forum)

SHELL

erected in outlying districts. Most of these houses have been constructed one storey high and many have been made demountable and others even mobile. A demonstration of the demountability of such construction was given in 1941 at India Head. Here, some hundreds of houses were erected for the workers of the U.S. Navy Powder Factory. The demonstration of demounting began at 7.30 a.m. and by 10.5 a.m., all sections had been dismantled. Loaded on four trucks the parts were taken one hour's distance away. Erection on a new site started immediately, and the house was re-erected complete by 2.30 p.m. A single storey type of house was selected, consisting of prefabricated timber frame sections covered with Homasote board.

The Precision-Built system of the Homasote Company of Trenton, New Jersey, is used mainly for bungalows and consists of complete wall sections rather than

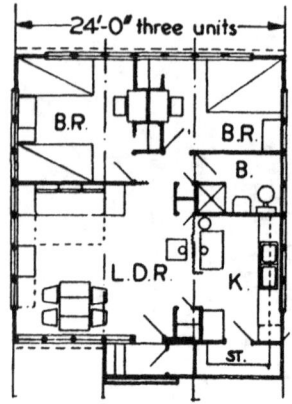

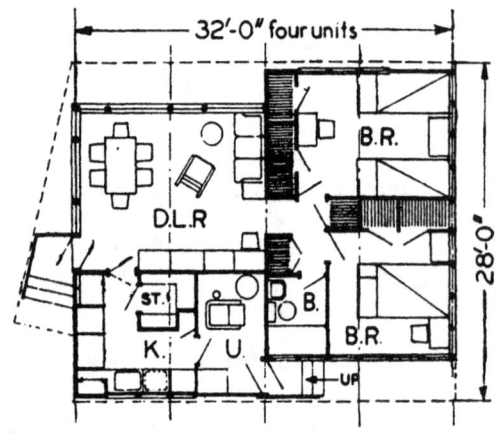

Plan of Tennessee Valley Authority demountable multi-unit houses. (See elevation on previous page)

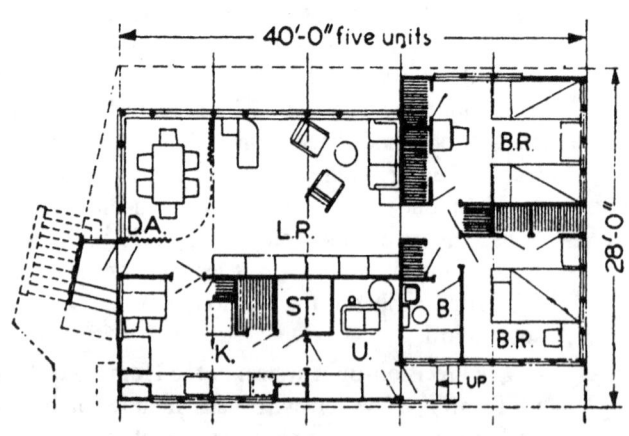

SHELL

SHELL

panels. The timber framing of these sections is covered on both sides with Homasote pulp board, glued to the face of the framing by means of special glue gums. Complete factory finish is given to these wall sections including the final decorating and the installation of conduits for electric wiring. Floor sections are put together on the site. Much research has been given to the evolution of this system, including the design of special transport and lifting devices for the handling of such large units.

Both the Gunnison Housing Company and the Douglas Fir Plywood Association of the U.S.A. have produced a large output of houses using $8' \times 4'$ panels instead of large sections. The D.F.P. Association face the panels externally with waterproof paper and battens

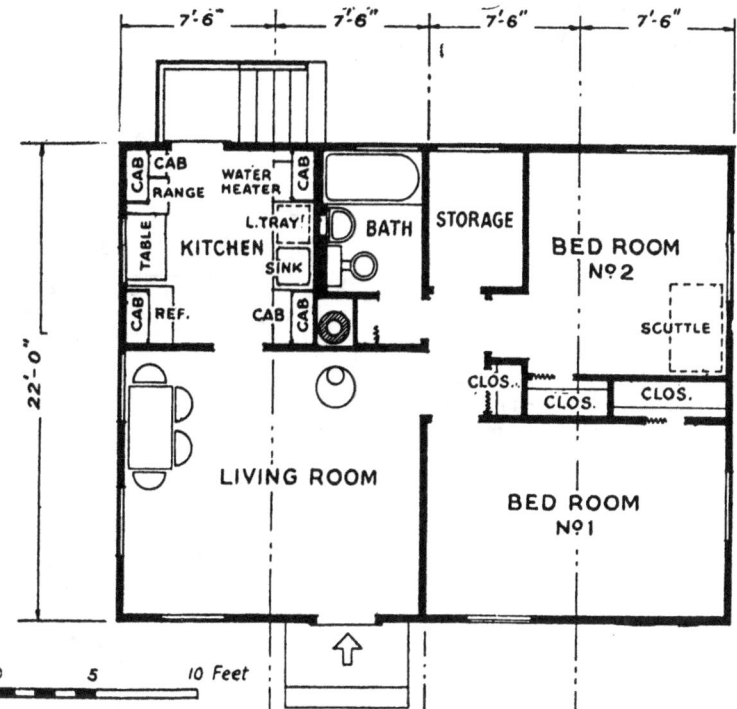

Plan of demountable two-bedroom type houses erected by the Tennessee Valley Authority for defence workers in the Muscle Shoals area of Northern Alabama. Elevations appear on next page

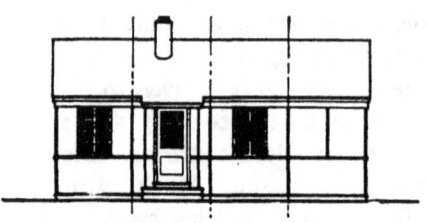

FRONT ELEVATION

LEFT SIDE ELEVATION

0　5　10　15　20 Feet

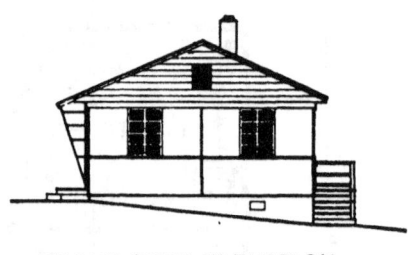

RIGHT SIDE ELEVATION

Tennessee Valley Authority demountable two-bedroom type houses. (Plan shown on previous page)

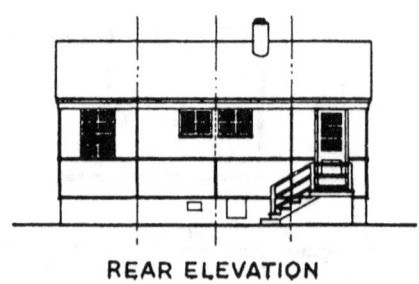

REAR ELEVATION

and finish with sidings, plywood or shingles. A special kind of splined joint is used at the junctions of the panels. A similar system is adopted by the Bates Prefabricated Structures of Oakland, California.

The prefabrication of U.S.A. wartime houses ranges between wall sections and complete dwellings. With the latter, the economic radius for distribution decreases as the size and weight of the unit product increases. A completely fabricated two-storey house such as the Van Ness, of Akron, Ohio, is purely local. Homasote, manufacturing sections of considerable size, always erect a factory near the site. National Homes, with more normal sections, operate up to a radius of three hundred miles. While the U.S.A. are already searching for a more permanent technique for post-war housing, their wartime experience of factory production is of the greatest value and, if we choose to profit by it, could be of immense service to this country.

We can also learn valuable lessons from Swedish experience of prefabricated timber house building. Soon after the first world war the need arose in Sweden to simplify the site erection of timber houses, so that unskilled future occupiers could erect their own homes. This led to the manufacture and design of fully finished wall and floor sections. In 1923, Forssjo, a Swedish timber business attempted the manufacture, for export to this country, of complete timber houses planned by English architects. These building units arrived in this country complete with roof tiles, window glass and door furniture, the parts being ready for assembly on the site and all nails and bolts provided. While the English public did not show undue prejudice towards the timber home, the parts when delivered still left too large a contribution to be made by the traditional builder in respect of foundations, chimney breasts, drainage, plumbing and other services. A further obstacle was the fact that the imported parts were based on the Continental metric system and could only with great difficulty be applied to the British workman's two-foot rule. Local bye-laws in built-up areas also necessitated an application for an annual licence for such construction to remain from year to year.

However, the system was an example of standardised and scientifically designed building units, far in advance of its time.

It is significant that Sweden to-day regards timber as a valuable national asset. This new value placed on timber has resulted in a scientific approach to its use and a cessation of any profligate waste. Sweden's timber is now finding profitable new markets in the form of viscose, cattle food, alcohol, and so forth, in addition to providing a valuable export trade. The country is therefore rapidly developing alternatives to the traditional timber house. Brickfields are being developed and light forms of concrete construction are being evolved for house-building purposes. It may be noted also that in the interests of timber economy, steel is being used for the heavier structural members. A recent example of this is a system of prefabricated timber and steel construction, designed by Eric Friberger. An interesting advantage of such a structural steel frame is that in a two-storey dwelling most of the ground floor can be left open as a covered space for the earlier life of the building, and this space may later be filled in with standardised wall units, when extension of accommodation is required.

SKELETON

LIGHT METAL ALLOYS

An experimental system which uses aluminium alloy as a structural skin has been evolved by the Aircraft Industries Research on Housing. A prototype design, one storey in height and based on the British Government's emergency house plan, has been found on erection to weigh only 1.9 tons. This system, which is really of the stressed skin type, is apparently adaptable to buildings of two storeys. The lightness of aluminium alloy is a considerable advantage in erecting the factory-made house, for it must reduce time needed for handling. An example of this was provided when the Rockefeller Centre was being built in New York. The wall spaces between the windows consisted of large panels of aluminium; and they were so designed that each panel could be picked up and put into place by two men.

WHAT METHODS CAN BE USED? 67

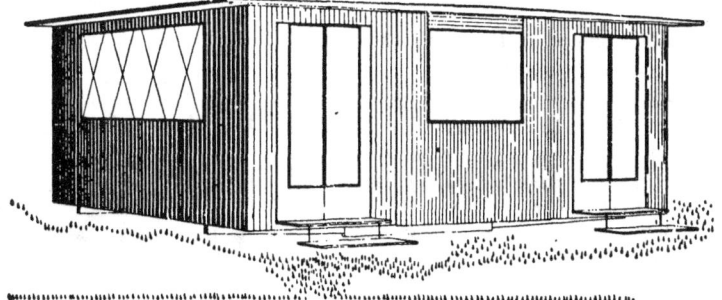

A trailer week-end house designed by R. D. Russell, R.D.I., for Aluminium Union Limited. Plan and sections, closed and open are given on the next two pages. To be extended, the trailer must be drawn on a fairly level site. Two telescopic cantilever girders, which are hinged in their length to hold up horizontally during transit, are let down to their vertical positions. On the underside of these girders is a hydraulically actuated self-levelling jacking system which is then put into operation so that the centre of the trailer is supported off the wheels and is level in both directions. The roof extensions are hinged along their top edges to the fixed centre roof and are lifted into position, being supported by folding stays. The telescopic girders are engaged in slots in the bottom of the main extending side walls (shown blacked in details on next page and on page 69) and a double-acting worm gear housed inside the girders is operated to extend them. The extending girders push out the main side walls—these are strengthened at the bottom by wide angle members to take the thrust—and the main side walls pull open the triple-hinged wall sections at back and front. During the whole operation the weight of the extension walls is carried by the cantilever girders. The roof extensions are then slightly lowered on to the walls and engage to form a tight joint. The floor extensions, which are hinged along their lower edges to the fixed centre floor, are then lowered into their horizontal position and engage with the bottom of the extension walls. The floor extensions are also carried on the cantilever girders and, between these girders, their thickness is increased to give the necessary strength. Where the large window openings occur in the main extension walls, the structural bracing members are taken uninterrupted through these openings to maintain the strength of the slab.

The outside sheets of the walls generally are of corrugated aluminium of about one inch pitch, the corrugations running vertically. All glazing is in transparent plastic sheet for the sake of lightness of weight and elasticity. The floor is covered with compressed cork tiles. The metal is anodised throughout as a protective treatment. The corrugated outside surfaces are clear olive green. The roof, including edge and soffit, all outside plain surfaces such as doors and window frames, and all metal inside are white.

F

KEY TO SECTIONS and PLANS

A. Curtains either shutting off whole extensions as bedrooms or enclosing only during day.
B. Bed. The frame is an aluminium tube on to which spring strips are mounted forming a shallow basket. During transit or when used as bunks, the moulded rubber mattresses fit inside these baskets, as beds, they are laid on top of the inverted frames.
C. Cupboard with tambour front containing calor gas cylinder, gas refrigerator and storage space. Cooking units and a small sink are in the top.
D. Doors to outside.
E. Chemical closet. The closed container is removable from outside.
F. Roof light.
G. Telescopic girder shown folded up in closed section and down in open section.
H. Hanging space for clothes.
J. Hydraulically actuated self-levelling jacking system.
K. Folding easy chair.
L. Ship-type hinged lavatory basin with storage cupboards over and beside.
M. Folding tables with tops hinged in two sections to stack in space beside wheel runs.
N. Nesting chair.
O. Compressed cork tiling.
P. Increased thickness in centre part of floor extension to give necessary strength.
Q. Sliding doors.
R. Curtain track built in flush with ceiling.
S. Glazed cupboard for cooking and eating equipment. The cupboard is backed by a window.
T. Trays for clothes.
V. Wheels.
W. Wheel runs with storage space beside.

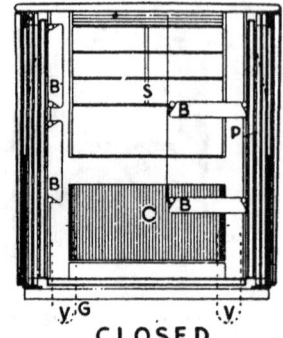

CLOSED

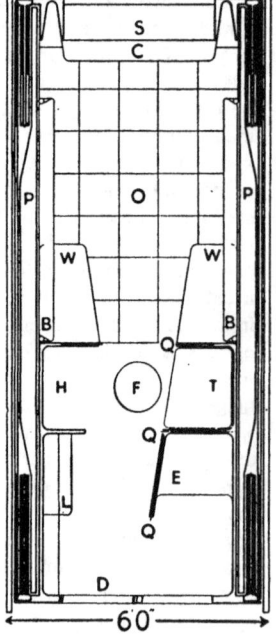

CLOSED

(These diagrams and the drawing on page 67 are reproduced by courtesy of Aluminium Union Limited.)

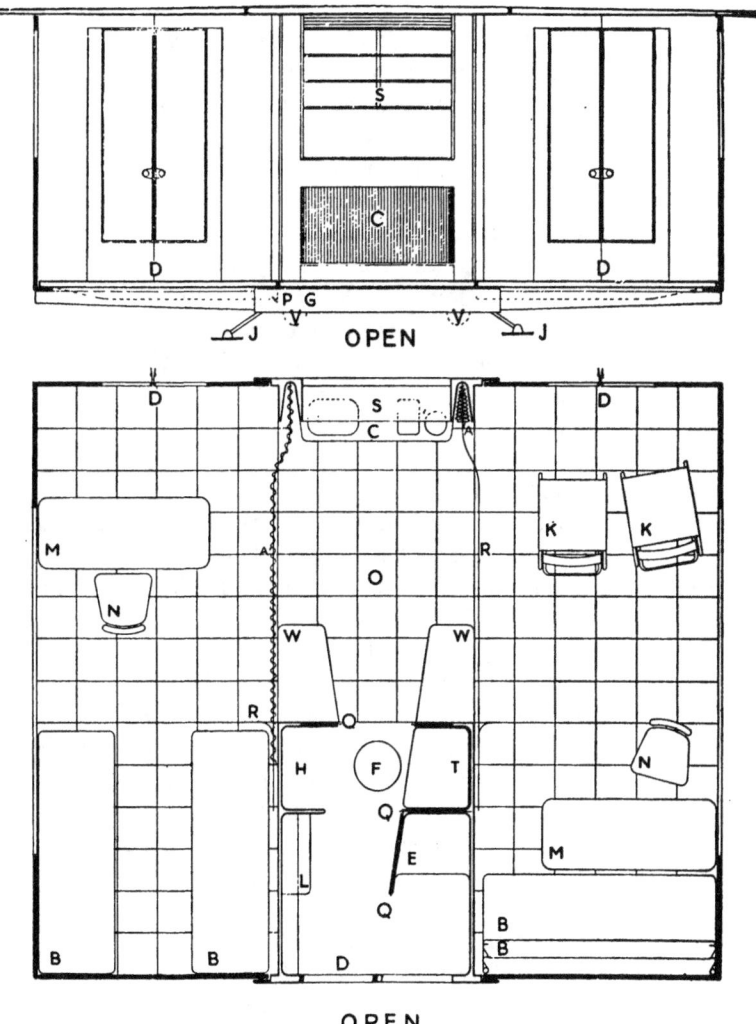

Section and plan of trailer week-end house, open; shown closed on previous page, with Key. (See page 67)

SHELL

SHELL

A mobile, expandable house has been designed by R. D. Russell, R.D.I., for Aluminium Union Limited. It can be used as a trailer, and unextended it is a compact and comfortable caravan. Set down on a temporary site, it expands to the dimensions of a small week-end bungalow. Externally, the walls are of corrugated aluminium, the 1″ corrugations running vertically. The metal surface is anodised. All windows have transparent plastic sheet, to keep down weight and increase elasticity. No other metal and few other combinations of materials could provide the spaciousness achieved by this expanding trailer week-end house, with the lightness of weight that enables it to be towed by a small car.

Light metal alloys, like plastics, will make many practical and economic contributions to composite systems.

CAST IRON

SHELL

Houses have been made of cast iron, and as recently as 1927 several houses were built on the Thorncliffe system, developed by Newton Chambers & Company, Limited. For these houses flanged plates of cast iron, 3′ square, were used, keyed on the face to receive a $\frac{3}{4}$″ outer skin of vibrated concrete. The flanges of the plates were bolted together, with an intervening wood fillet to accommodate the fixing of wallboard lining on the interior. These coated plates weighed two cwt. each, and were heavy to handle.

SHELL

An early experiment in cast iron house design was No. 1 Lockhouse, at Tipton Green, Staffordshire. This was a bungalow, and was apparently built late in the eighteenth or early in the nineteenth century. A few years before it was pulled down it was fully described in an article published in the *Birmingham Gazette*, December 6, 1924. It was there stated that " The iron plates of the walls are flanged, and it is apparent that the flanges are used, in the interior, to take the wooden laths that support the plaster. The effect is that there is a cavity between the plaster walls and the iron walls—a circumstance which possibly explains the dry and comfortable character of the house."

COMPOSITE

In the ten years 1929-1939, a great deal of significant research work was done in the United States, and developments of the greatest importance resulted. Research was undertaken by the Beamis, Pierce and Purdue Foundations, and by various agencies set up by the Federal Government. This work by organisations interested in low-cost and better housing has proved invaluable. The creation of a Federal Housing Administration made possible, too, the private financing of low-cost housing for persons of limited means and established minimum standards of space, equipment and construction. This provided a yardstick for evaluation. Not the least important during this period was the activity of design research workers, who kept

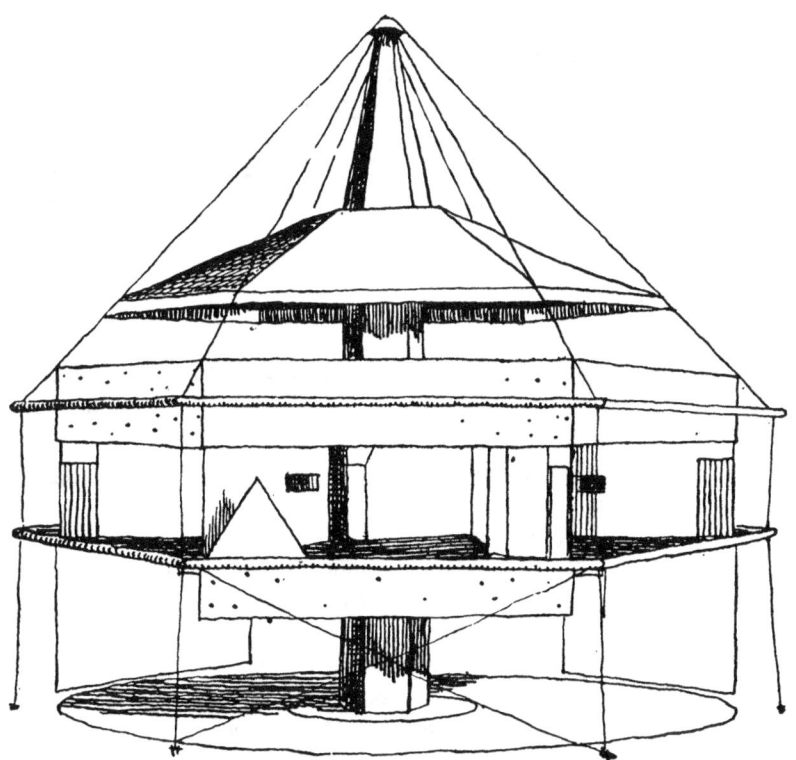

The mast-hung Dymaxion house designed by R. Buckminster Fuller
(See pages 72, 109 to 111)

HOUSE OUT OF FACTORY

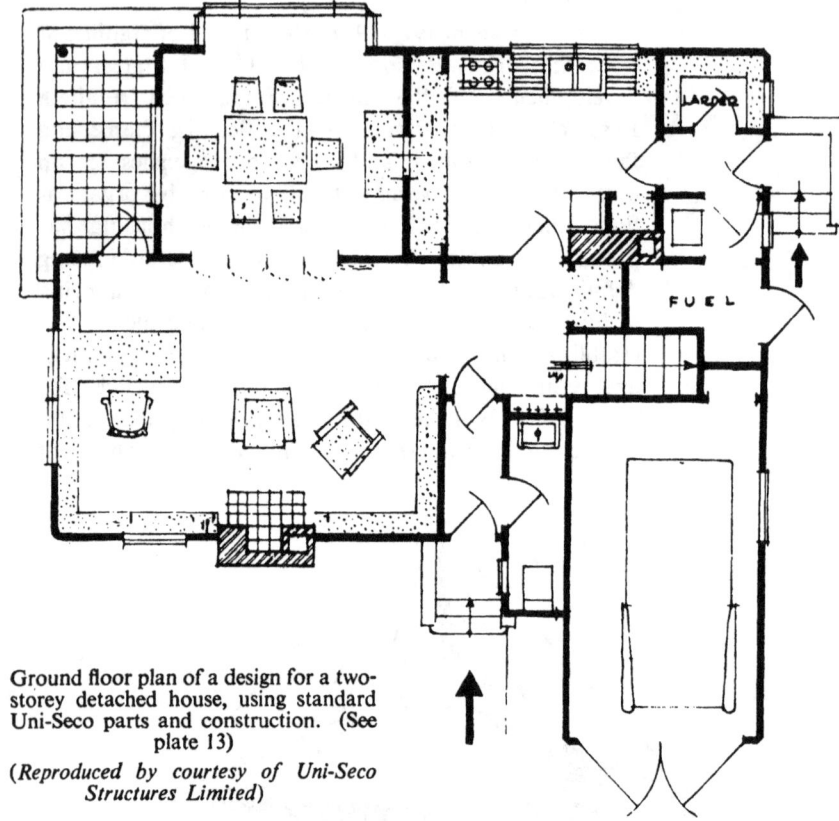

Ground floor plan of a design for a two-storey detached house, using standard Uni-Seco parts and construction. (See plate 13)
(*Reproduced by courtesy of Uni-Seco Structures Limited*)

the public, through the Press, supplied with new ideas which promised to reduce household drudgery by more and better mass-produced equipment.

These various ideas suggested three new kinds of habitation:
1. The mast-hung house;
2. The eggshell or monocoque house (such as a trailer or half sphere);
3. The machine-core house, which included as a self-contained unit (like a car engine) all the mechanical services of the home.

These ideas were put into model or demonstration form; Mr. Buckminster Fuller's mast-hung Dymaxion house (1929) and later his four-piece bathroom attaining immense and unforgettable publicity. (See page 71).

WHAT METHODS CAN BE USED? 73

The whole approach to house design was changed by these ideas; and the conception of the factory-made house became detached from traditional modes of architectural thought. It could now be identified as an industrial product; it was not, nor could it ever be again, regarded as architecture. But even these new revolutionary methods of construction could be classified as *Shell* or *Skeleton* systems. They used a variety of materials in association.

In Britain some composite systems were developed during the war period. One system originally evolved for the purpose of war building is called the Uni-Seco.

SHELL It is based on the use of standard panels about 7' high by 3' wide by 1¾" thick. These consist of light wood frames into which are grooved on both sides asbestos cement sheets filled with wood wool and cement.

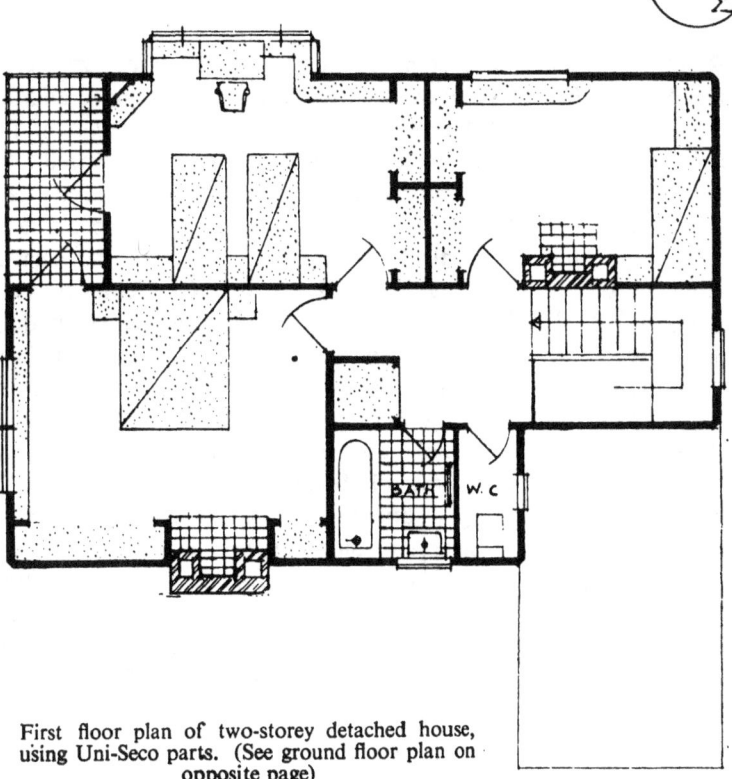

First floor plan of two-storey detached house, using Uni-Seco parts. (See ground floor plan on opposite page)

This filling, while wet, adheres to the asbestos cement sheet, considerably strengthening it and reducing brittleness. These panels are joined with wood tongues and then screwed together. A special mastic filling is put on the outside. They are completely load-bearing. Floors are made in long panels extending from wall to wall, 3' in width and 3" deep, with plywood glued both sides to bearers between them, thus constituting a stressed skin construction. The span is 12' 9". They weigh approximately 3 lb. per foot. When laying these floor units a space of 4" is left for service pipes, and then covered over. Beams or lintels are placed at the level of the ground, the first floor and the eaves to tie the ends of the panels together. Beams of hollow box sections made of plywood sheets glued together with synthetic resin adhesives are used to support the roof panels, so saving timber and weight.

Research continues, and some of the most distinguished architects and industrial designers are engaged upon it and have made and are making significant contributions to the design and character of the factory-made house. Although his work has not been specifically concerned with housing, the American industrial designer, Raymond Loewy, has made many important improvements to the technique of prefabricated building, particularly in connection with the smaller stations of the Pennsylvania Railroad system. Of architects, now working in America, Walter Gropius has an international reputation, and is the best known of all the pioneers who have studied the application of mass-production principles to building practice. He established his great reputation on the completion of the Bauhaus at Dessau in 1926, and his activities have continued and his influence has increased ever since. He has recently become associated in the U.S.A. with a " Packaged " system of building. This system uses light wood standardised frames for structural purposes, taking a metal clip and wooden wedge jointing component, which allows considerable variation in the arrangement of the frames. As the frames are left four square, without any fixing projections, they can easily be packaged and man-handled, and after erection they can take various types of covering.[1]

---

[1] This system is described and fully illustrated in *The New Pencil Points*, April, 1943.

WHAT METHODS CAN BE USED? 75

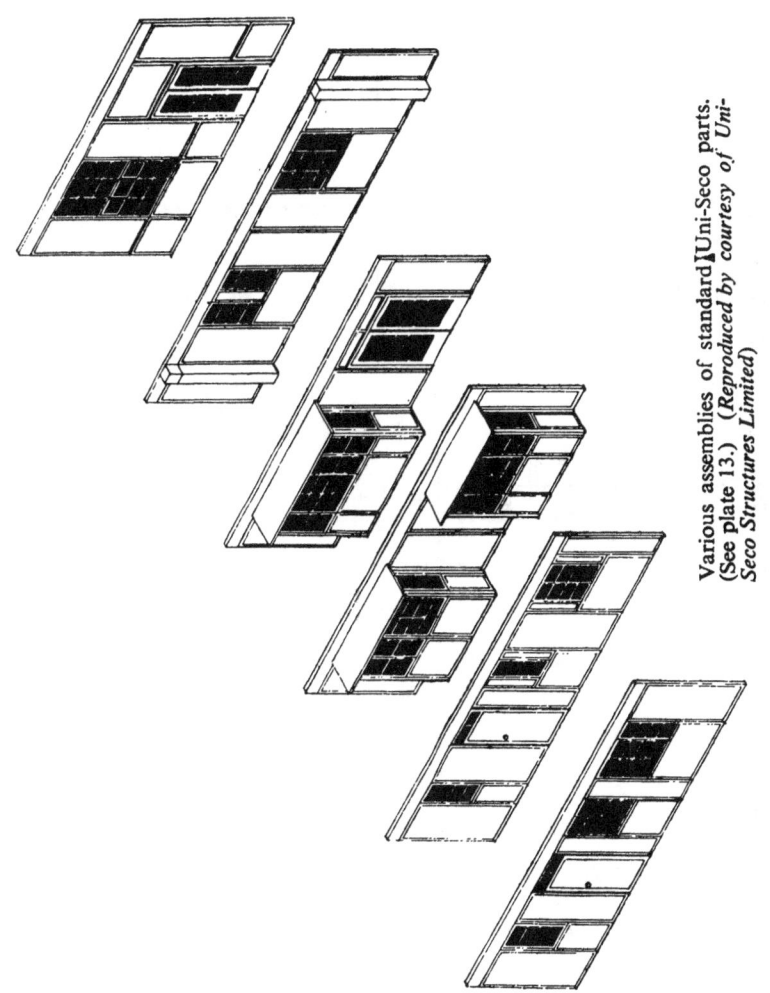

Various assemblies of standard Uni-Seco parts. (See plate 13.) *(Reproduced by courtesy of Uni-Seco Structures Limited)*

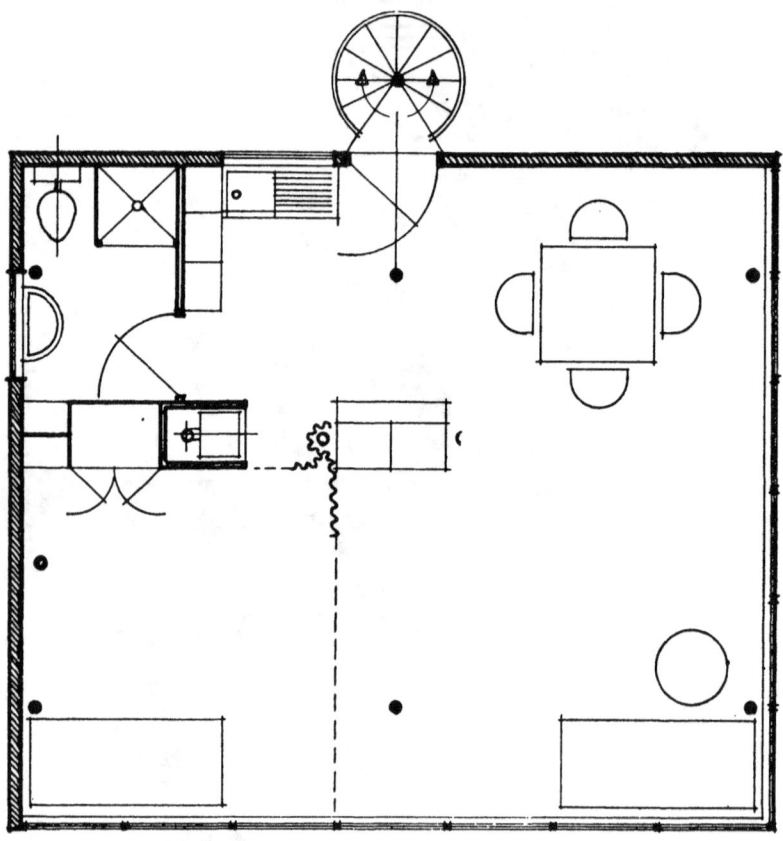

Plan of holiday house at Northport, Long Island, N.Y., U.S.A. (See opposite)

WHAT METHODS CAN BE USED?

Two views of a holiday house at Northport, Long Island, N.Y., U.S.A., designed by Kocher and Frey. Plan shown opposite

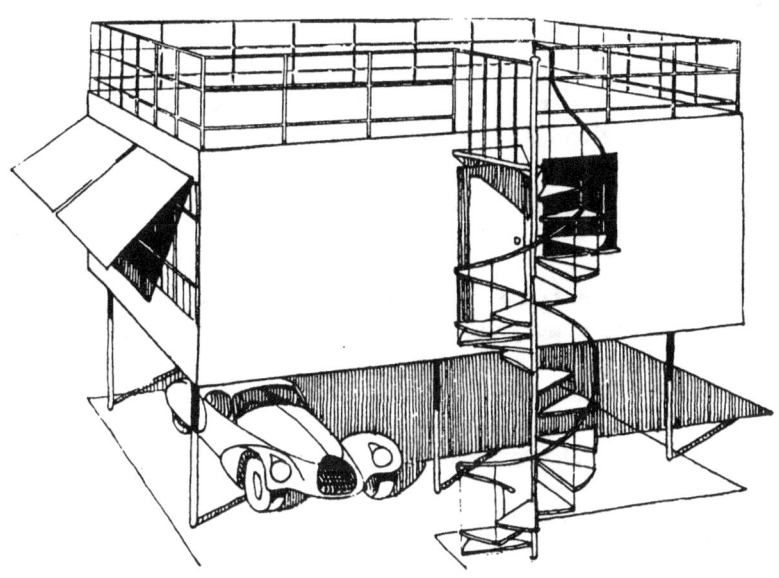

A. Lawrence Kocher and Albert Frey have designed the Aluminaire house; a structure with a light steel and aluminium frame. It was first erected in 1931 at an exhibition in New York and re-erected at Long Island. They have followed this up with many further experiments in light frame construction; and in 1934 they built at Northport, Long Island, a house of exceptional originality. It is a " canvas house," with its single storey raised upon steel stilts above the level of the surrounding woodland. A spiral staircase gives access to the house. It has been described in some detail as follows:

" Actually it is a compact, comfortable and good-looking week-end house. A one-storey structure, it provides two living-storeys and a garage. It is supported by six light steel columns, embedded in concrete at the base. The space below the house accommodates two cars; the principal floor, which is the only enclosed space, is subdivided into living- and dining-room, kitchen, bathroom and store. At night the living-room may be sub-divided by curtains into three bedrooms, all with access to the bathroom. The roof terrace, the third storey, is an outdoor sitting-place by day and a summer-time bedroom. The windows are of the steel projecting type, with glass that admits ultra violet rays, and they are orientated to secure views of water and woodland. The entire exposed face of the house is of painted canvas duck, stretched over a grooved-and-tongued facing of redwood flooring. The walls are insulated by paper-backed aluminium foil placed as a continuous membrane between the exterior and the interior of the 4" wall. The interior wall facing is of $\frac{1}{4}$" plywood. Curtains exclude direct sunlight on hot days, and awnings help to shield the windows."[1]

O. Kleine Fulmer designed, in 1933, a two-storey frameless steel house, using 18 and 20 gauge hot rolled sheet steel, pressed into box-like channels and other sections, for walls, floors and roof. Welding was used extensively and large sections were thus assembled for erection. The exterior was faced with porcelain enamelled metal sheets.

Richard J. Neutra has contributed much to evolving new formulae for the modern house, particularly in the use of light steel construction.

From all these experiments in design a new conception of the dwelling is arising. How well this conception could be handled by industry and how the ultimate results may change our lives are subjects examined in the chapters that follow.

---

[1] *The Book of the Modern House*, edited by Sir Patrick Abercrombie. Chapter XI, " The House in America: A Comparison," by John Gloag. (Hodder & Stoughton, 1939.)

CHAPTER 4

Can Homes be Mass-produced?

HAVING examined the materials and the constructional principles used for various prefabricating systems, it is necessary to appreciate the advantages and possibilities of factory production. It cannot be denied that in Britain we have inherited a lot of prejudices from the early days of industry, when factories were engaged in turning out cheap and nasty versions of articles that had once been made conscientiously and skilfully by craftsmen. For nearly a century mass-production was largely dedicated to the task of making poorly conceived objects look rich. All kinds of new processes and materials were deliberately employed to hide their industrial origin, and few manufacturers had the wit to design the things they produced so as to make the best use of the industrial technique which they had mastered. It was not until such things as motor cars, radio receiving sets, electric light bulbs, tooth-brushes, and refrigerators, were produced to make the best possible use of materials and manufacturing methods that many people (including manufacturers and distributors) began to realise that mass-production was not an operation inevitably tainted with vulgarity.

Although nobody would object to the convenience of a mass-produced motor car; although nobody would dream of saying that Henry Ford or Lord Nuffield had by their manufacturing activity imposed a monotonous pattern of life on the peoples of the United States and Great Britain, when precisely the same principles of production are applied to the making of houses every old-world prejudice is aroused in Britain. The factory-built house is opposed on the grounds that it is a machine-made job, and in Britain we have never accepted the machine with the enthusiasm with which it has been accepted for generations in the United States. America's huge westward expansion in the nineteenth century was made possible, first by the steamboat and then by the railroad. Americans took to the telephone and the typewriter years before we did; Henry Ford was the first industrialist to think of the motor car as a universal convenience and not as a rich man's toy. Consequently, there is comparatively little " consumer resistance " to the idea of mass-produced, factory-made houses in the United States ; but here the thought of making a home in such a place repels many otherwise quite intelligent people.

The popularity among highbrow architects and social reformers of M. le Corbusier's slick phrase, " A house is a machine for living in," has been unfortunate. But it has never been popular with the public, nor could it have ever become acceptable save among people who ignored the humour, the intense individualism, and the rich home-making capacity of the majority of their fellow-countrymen. Perhaps that facile phrase has led many people to feel that a machine-produced house must necessarily be " a machine for living in." It need be nothing of the kind, though it could be as convenient and efficient for its purpose as a well designed motor car is convenient and efficient. In car production, engine, chassis and body are built up on the assembly line, coming together at the appropriate time, so that at the end of the line the car can be driven away on test. An analogous procedure allows the various parts of the house to be made in the factory for assembly on the site. We accept the car as a modern convenience ; we accept the motor bus and the trolley bus, which are factory-produced ; we don't clamour for a cabriolet or a stage-coach ; though when mass-production is suggested for house making we clamour for methods and materials that modern industrial technique has rendered partly obsolete. It is a fact that the application of mass-production to house building has in the past been tentative : an infiltration of new ideas has, from time to time, furnished a considerable variety of standardised house parts, progressively convenient and, if used in sufficient quantities, comfortingly economical. But hitherto standardised window frames, doors, plumbing units, kitchen equipment, and so forth have been thought of as separate items, unrelated by any common system of dimensions which would enable them to be assembled with the maximum of speed and the minimum of labour.

If we pursue the analogy of car production we come to the machine conception of house production, which is vastly different from the arid doctrine that the house itself is " a machine for living in." This machine conception is perfectly straightforward commonsense. The mechanical parts of the house, the plumbing, heating and cooking units, are regarded as the *engine*. The house is the *body* which is activated by this engine. If the analogy appears outrageous to those who believe, as Wren believed, that "Building certainly ought to have the Attribute of Eternal," they should remember that Wren said that in a century with a different social and economic system from ours, that industrial technique was not then available and, had it been, a genius like Wren would unquestionably have used it to superb advantage. But whether the technique of mass-production will be used to advantage to-day depends upon who designs the units, who directs the assembly, determines the finish and lays out the site. If the architect directs the making of houses from mass-produced, prefabricated " engine " and " body " units we shall have design ; if the speculative builder gets to work we may only have disguise. Seldom have architects in Britain or America had such an opportunity—or such a responsibility. Seldom, too, has the British public had such a prospect of comfort and convenience in the home. So far the ideas of the public, their tastes and standards, have been dominated almost entirely by the speculative builder. He has sold his goods on the appeal of price, on superficial stunts, on picturesque appearance. Such terms as " labour-saving," " artistic," " bijou," and even " architect-built " have become the commonplace material of high-powered salesmanship : their meaning varies. A few tiles over the draining boards and sink may mean that an ill-planned, cramped kitchen is labelled " labour-saving :" a roof with drooping eaves or a front door of stained oak studded with nails may be " artistic." " Bijou " may mean anything.

Before discussing the economic and technical advantages of using the technique of mass-production for house building it will be illuminating to glance, briefly, at the results of using traditional methods in the interval between the two wars. During that time the low grade jerry builder managed to suggest that the Englishman's castle was really a cottage ; and the rows and rows of semi-Tudor, semi-Swiss gabled " villas," with their stained lath and stucco " half-timbering," and their casements and front doors with leaded lights in the upper panels, represent the vernacular architecture of England. These houses provide the illusion of privacy ; though as work is often scamped on them, and little use is made of adequate insulating materials, nobody can be private in them, since the loud speaker can distribute impartially to the world at large the vapourings of politicians, the oratory of statesmen, or the humours of Tommy Handley. Bricks, mortar, tiles, the cheapest woodwork, plumbing that is just plausible, and electric installations with points inconveniently placed to provide pinprick annoyances in perpetuity—all these casually assembled materials and services are accepted because they convey an air of spurious permanence. Put together as a house or bungalow, they may even outlast by a few years the paying of the final instalment of the purchase price.

This vernacular domestic architecture of Britain, produced by the speculative builder, is thus sustained by two illusions : Privacy and Permanence. It should be noted that it is produced largely for the consumption of the middle classes, and that the working classes are, generally speaking, better housed in terms of architectural design, for it may be repeated *that Britain led the world between the two wars in slum-clearance, and the laying out of housing estates and great residential blocks for the occupation of industrial workers.* These working-class housing schemes were often erected upon the best technical advice, and made good use of modern building methods and materials. The standards set by the architects who designed them were lowered only when the municipalities who furnished the finance were startled by expense, and curtailed some obvious convenience. (There are some archaic people, still in positions of authority in this country, who believe that every working-class household will automatically confuse the bathroom with the coal cellar.)

Now the factory-made house has been considered primarily as a domicile for the workers : the middle classes, who represent a not unimportant section of the population, appear to be ignored in the housing plans made by the appropriate state departments. But most of the criticism of factory-made houses has come from people who may never have to live in them. They are to be available only for the use of the working-classes. Middle-class critics have used rather unrealistic standards when passing judgment unfavourably upon factory-built houses. They ignore or at least minimise the convenient equipment of such houses, and they rest their objections largely upon aesthetic grounds. They fear the monotony these houses may impose upon the rural or suburban scene, forgetting the monotony that bricks and mortar, in unskilled hands, can inflict upon any locality. They doubt whether it will be possible to be private in a factory-built house ; they forget the aural promiscuity of a row of semi-detached speculatively built houses, or the terraces of a housing scheme, or the apartment blocks of a slum clearance scheme. Although it is impossible to preserve the illusion in real life that bricks and mortar, when carelessly and cheaply handled, guarantee privacy, the criticism is frequently heard that the factory-made house " isn't solid," that it is " flimsy " or " tinny." Life in it, some boisterous critics assert, will be like life in a tin can, and a householder doesn't want a tin opener for his house, but a real door key, and a real solid cosy house behind the front door. Of course, such views arise from the old-fashioned notion that anything mass-produced is cheap, nasty and vulgar. But what are the technical and economic facts about mass-production in relation to houses ?

There are, as we have seen, many systems for turning houses out of factories in the form of easily assembled units and equipment, and most of these systems permit the most practical use to be made of existing knowledge of modern materials, modern plumbing, ventilation and lighting, heating and insulation. Although during the last thirty years, the most important discoveries and advances have been made in the science of insulating enclosed spaces against the penetration of sound waves, that technical knowledge is seldom used in Britain. Since broadcasting has developed, the importance of applying such knowledge has greatly increased; and modern traffic, both by road and air, has made it imperative to protect the householder against a round-the-clock assault upon his ears. Bricks and mortar, lath and plaster—all those traditional materials—should be accompanied by the use of sound-absorbing materials; but they seldom are, because the jerry-builder builds cheaply to clear the biggest possible profit, and the housing schemes sponsored by municipalities are frequently cut down to the bare minimum because of cost; therefore *the finance is never available for giving to householders of all classes the benefits that modern technical knowledge could confer upon their homes.*

Now the production of houses in a factory, the pouring out of prefabricated units and equipment for transport to the place of assembly, would speed up and might cheapen the business of house building. The money which can never be spared for proper insulation and good domestic equipment, such as conveniently disposed stoves, storage units, sinks, refrigerators and washing-machines, would at last be found; not, as the social reformers believe, because the State should provide it, but because *the economic laws of mass-production furnish an expendible surplus as a result of savings in time and labour* which will become available for the amenities that are technically possible but are always " extras " when traditional methods of building are used.

The factory-made house, then, could have better amenities than the ordinary traditionally built house, in a speculative builder's or a municipal scheme. It could have greater privacy, and because it is better insulated, not only against sound, but also against heat and cold, its running costs would be lower, for less fuel would be needed to keep it warm in winter.

If we return, for a moment, to the analogy of motor car production, it will be realised that the designing and manufacturing of a convenient and truly labour-saving house, with its " machine parts," its cooking, heating and plumbing units, mass-produced and easily assembled, allow a careful and imaginative study of consumer needs to be made. In the motor car industry such study is normal industrial procedure ; without it, no car would have a chance of acceptance, and the car using public is acutely critical of " performance " and the consumption of petrol, as well as of appearance and general efficiency. How many potential home-makers are correspondingly alert about the " performance " a house will give, and how much fuel and power it will consume in cooking and heating ?

The " performance " of a house can be tackled as a problem of design, and the way designers set about the problem is particularly instructive. As an example, we give the following highly condensed history of the experience gained by a big American housing corporation which has developed prefabricated housing on a large scale.[1] This organisation, the Gunnison Housing Corporation, has contributed in a big way towards the defence housing programme and is now one of the largest of its kind in the U.S.A. Foster Gunnison, its founder, was trained originally to be a journalist, but never followed that profession ; instead he acquired a small lighting business, which in a few years was turned by his enterprise and energy into a highly successful concern. He abandoned that business and, backed by Owen D. Young, of the General Electric Company, began experimental design research work on prefabricated houses, acting as the co-ordinator of a team of specialists. " After four years they produced a house, but it was too like a laboratory product and although it excited a lot of comment no one wanted to buy it. It was then that Gunnison realised that they had been looking at the problem too much from the technical point of view and had been trying to adjust house design to machines instead of the other way round."[2] Gunnison started again, and his backer gave him a free hand to use any materials or equipment. After two years he had produced a house that could become a home. " The basic structural panel is timber-framed, packed with mineral wood insulation and covered both sides with sheets of plastic and resin bonded plywood. The firm now makes nine standard sizes of house in two price ranges and provides ' architectural features ' which can be added to taste, which may include a garage, a porch, window boxes, paths, drives and fencing."[3] The Gunnison Housing Corporation has organised departments to cover research, engineering, production, erection, sales, advertising, mortgage financing, estate management, legal matters and accounts. The plan and equipment of a house

---

[1] A more detailed account of the work of this Corporation was given in an article entitled " Prefabrication in America," published in the *Architects' Journal*, Vol. 96. No. 2489, October 8, 1942, pp. 231—233,

[2] The *Architects' Journal*, October 8, 1942, p. 231.

[3] *Ibid*, p. 231.

are determined by the Sales Department, which is in touch with consumer needs.

Americans are accustomed to, and indeed demand, a much higher standard of domestic equipment than the British public. Consequently, the American housewife has less drudgery and more leisure. Drudgery can be eliminated and leisure increased by competent designers : the necessary inventions have been made, the devices exist ; and American makers of houses would no more dream of producing houses without them than would a motor car manufacturer of producing cars without four-wheel brakes and self-starters. By turning a tap or a switch, pressing a button or moving a lever, the American housewife reduces not only the time taken up by housework, but escapes the harassing exasperation of coping with tasks which are totally unnecessary in this machine age. It is the job of an efficient sales department in an American organisation concerned with housing, not only to know what the housewife expects, but to give her something better than she expects. With competent industrial designers and all the powers of industrial production behind them, such organisations can progressively improve the factory-built house and its equipment : it is good business to give the public something ahead of their expectations. In our country, the speculative builder has a totally different outlook : his policy is to give the public the minimum they will accept, and to doll it up to look picturesque and make it easy to rent or buy. Municipal authorities also share this obsession with minimum standards.

The mass-production of houses in America has proved that we could change over to maximum standards in every type of factory-built house. The competition from such houses would force the speculative builder to go out of business or to bring his ideas up to date : either course would be of enduring benefit to the British public.

The money can be found for maximum standards : it can be found in the factory. We can choose high rates, high standards and old-fashioned building methods, or low rates, low standards, and still cling to tradition ; or we can at last acknowledge that we are living in a commercial machine age, have maximum standards, and pay for them by letting the machine *save the time that is money* in the building of houses.

CHAPTER 5

How Long Will it Last?

Houses that endure for generations often become tyrannical owners of those who inhabit them. The old house, with its charm (or lack of it), its spacious air, its inconveniences, its patch-work application of modern appliances, constantly added to so that the householder may be able to live appropriately in his own century, gradually takes control of the life that goes on within its walls. It imposes restrictions; and the fact that many of those restrictions are cheerfully accepted because the external features of the house are beautiful, does not remove although it may mitigate their cramping effect upon life. Only when an old house is fairly large can its advantages wholly outweigh its disadvantages; the converted cottage is often a nuisance, unless it is enlarged. Life cannot be based only upon the possession of a few modern heating, cooking and refrigerating appliances: the dwelling that accommodates such things must include the advantages that modern methods of construction and insulation and good planning can bring. But people are often compelled to live, without modern appliances, in part of an old house that has been converted for the use of perhaps five families when it was originally designed for one; and such dwellings have no pretensions to modernity in planning or insulation. They are merely covered spaces for the reception of as many rent-paying human beings as possible, for one idea and one only inspires those who own and prepare them for habitation: greed.

Because houses last so much longer than people, and because many of the discarded houses of the nobility and gentry of former times are converted into slums, the life of large-scale traditional buildings often ends in decay and misery. This is not a social and economic evil peculiar to the last hundred years; it has haunted large cities for centuries. For example, in London, many of the majestic palaces that had lined the Strand in mediæval and Elizabethan times had degenerated sadly by the beginning of the eighteenth century; but the evil of over-crowding was far older. A proclamation issued by Queen Elizabeth in 1580 had urged her subjects " to forbear from letting or setting, or suffering any more families than one only to be placed, or to inhabit from henceforth in any house that heretofore hath been inhabited."

The conversion of a large house into apartments designed to accommodate several families only creates slums, when those houses have outlived their strength, and no steps are taken by unscrupulous landlords to regenerate them. Thousands of large, Victorian houses, built to accommodate the large families of that period, are now divided into self-contained apartments, each having its own bathroom and kitchen and heating appliances, for the day of the large town house has passed, and the day of the large country house is passing too. There may, in fact, be a period before the housing shortage in Britain is satisfied by industry when innumerable families will be billeted all over the country, as evacuees from bomb-threatened cities were billeted during the war, in large houses which cannot be fully occupied by their owners. All such converted houses, no matter with what skill they are converted to a use for which they were not originally designed, will impose their character upon the inhabitants. People living under such conditions are not really home-making: they are camping out, and complete privacy, so far as sound is concerned, can never be guaranteed under such circumstances. For a period, then, it may be essential to use large houses that have survived from the past ; but this should not be regarded as a permanent necessity.

One of the characters in Nathanial Hawthorne's book, *The House of the Seven Gables*, complains bitterly about living " in dead men's houses." This character, Holgrave, who is an artist, looks forward to the day " when no man shall build his house for posterity," and then he explains the application of his view to life. The book was written in 1851, but the views expressed are appropriate enough to-day. " If each generation were allowed and expected to build its own houses, that single change, comparatively unimportant in itself, would imply almost every reform which society is now suffering for. I doubt whether even our public edifices, our capitals, state houses, courthouses, city halls, and churches—should be built of such permanent materials as stone or brick. It were better that they should crumble to ruin once in twenty years, or thereabouts, as a hint to the people to examine into and reform the institutions which they symbolise." (Chapter XII.)

Fifty years later, H. G. Wells was putting a footnote to the third chapter of his *Anticipations*, in order to say that he found it incredible to think that there would not be a sweeping revolution in building methods during the next century. " The erection of a house wall, come to think of it," he said, " is an astonishingly tedious and complex business ; the final result exceedingly unsatisfactory." After a detailed and quite unanswerable criticism of building methods with brick, he discusses prefabrication, after saying, " I do not see at all why the walls of small dwelling houses should be so solid as they are. There still hangs about us the monumental traditions of the Pryamids. It should be possible to build sound, durable, and habitable houses of felted wire-netting and weather-proofed paper upon a light framework. This sort of thing is, no doubt, abominably ugly at present, but that is because architects and designers, being for the most part inordinately cultured and quite uneducated, are unable to cope with these fundamentally novel problems. A few energetic men might at any time set out to alter all this."

They could ; particularly with the results of fifty years of jerry building to warn them, and the need for millions of new homes to encourage their activity. The sort of materials which are available to-day for building go far beyond the tentative suggestions made by H. G. Wells ; and if such materials are used for making houses of good design, if they are properly handled and protected, then they could last at least as long as a jerry built brick house, and could be easier to live in and better to look at. There are some catch-phrases which have confused a good many issues in industrial production, and " cheap and nasty " is one of them. Cheapness does not imply nastiness. It costs no more for a factory-made house to be well designed.

The whole question of design tends to be confused by the belief, at first accepted by politicians, publicists and the public, that the factory-made house must necessarily be a temporary structure. Because of this, it was often assumed by people who ought to have known better that its design was comparatively unimportant ; it was *temporary*. It was only intended to last ten or twelve years ; it was implied that its appearance hardly mattered. But the factory-made house is not, nor should it ever have been regarded as, a temporary expedient : it is a house " in its own right," capable of lasting fifty years if necessary ; but it is also an industrial product.

In selecting and assembling the materials for the factory-made house, the matter of primary importance is design. Percy Smith, R.D.I., once said, " I would suggest that it is quality of design, even more than quality of material, which gives sustained value and interest. Bronze, for instance, is a noble material, but if something made of bronze is poorly designed it is boring instead of stimulating, and is soon neglected or used merely as a convenience and only until something better is produced." (*Journal of the Royal Society of Arts*, Vol. XCI, No. 4644, page 470.) In traditional building, many fine and long-lasting materials are employed, such as stone, brick and concrete. But for all their strength, if they are wrongly put together they can admit damp, and thereby rot timber, plaster and other materials in a building, and shorten its life and their own.

Even a leaky pipe can cause serious harm in the most substantially built dwelling. Timber is long lasting if kept properly dry and ventilated, and is thereby not subjected to dry rot. Roof tiles, though in themselves capable of lasting for hundreds of years, are dependent for life on the nails that fix them, and on the wooden battens that hold the nails. The material of metal flashings may last for years, but their efficiency is dependent on freedom from damage or displacement. The degree of permanence of buildings constructed with traditional materials is determined by their weakest parts.

The factory-made house may be designed to have either a consistent period of life for all its parts or a durable constructional frame, which would allow for the renewal of its more short-lived elements by using standardised factory-produced parts, such as outside covering units or portions of equipment. The life of the latter type of house might thus be extended by the renewal of parts, to, say fifty years.

Very short-lived factory-made dwellings are not considered in any detail in this book, for, in the view of the authors, they are not economically sound and they come into the category of the caravan dwellings which were used in the United States in the early nineteen forties, to alleviate a severe shortage of accommodation for war workers. For example, the Portal type of house had an intended use of ten years. Such houses could easily be designed to last twice that time, if the technique of having a durable constructional frame were adopted, so that the periodic renewal of fast-wearing parts might take place.

Building during wartime in Britain has shown that it is not possible to provide buildings which are both satisfactory when erected, and structurally short-lived.

Factory-made houses, well designed and using to the best advantage all the powers of mass-production and modern materials, are in less danger of social obsolescence than traditional buildings. While the short-lived factory-made house, which is produced merely as a stopgap, may after it has stood for twenty years or more be a new and most repellent type of slum, the flexibility of the well designed factory-made house provides a safeguard against the inevitable obsolescence that awaits the temporary dwelling and the traditionally built house. The factory-made house allows the internal alteration of the plan, it permits the renewal and improvement of all internal equipment and installations; and this is of particular importance with insulating linings. It may well be equally important in the matter of services and equipment, particularly the service of electricity for lighting, heating and cooking.

The initial design of the house could also enable extensions to be made, to accommodate additions to the occupying family. It should be just as possible for the small householder to add a new wing, containing a couple of extra rooms, to his factory-made house as it was for the eighteenth century nobleman to add a new wing to his mansion, containing ten or twenty bedrooms. Adding to a traditionally built house is a costly business : adding to a factory-made house could be relatively simple : the unit construction of such houses making it a matter of the greatest mechanical ease, and the additional units necessary for the extension could be erected in a few hours after they had been ordered from the factory.

The factory-made house could certainly last well, be easily brought up to date, and, because it could be expanded without much trouble, would never cramp the lives of its occupants.

CHAPTER 6

How Much Will it Cost?

EVERYTHING, we are told, is decided by price. In order to give a scientific flavour to the statement, some people add the phrase : " In the final analysis." Other people, with a liking for directness, say that when it comes down to brass tacks price is the thing that matters. Upon the fundamental significance of price there is almost universal agreement. Politicians shake their heads and admit it ; reforming idealists shake their fists and admit it ; while all those people concerned with traditional building accept it as a matter of course. The results of this widespread agreement vary ; but nearly everyone is unhelpfully negative whenever suggestions are made for higher standards and longer life for houses. The politicians become plaintive (" It's a question of what the country can afford ") ; the reformers indignant (" vested interests and privilege are conspiring to rob the workers of higher standards ") ; while the people who do the building get on with the job and make it cheap, and, because they know no better, they often make it nasty (" people just want *houses*, and they'll take what they can get ").

Working *down* to a price can be a stimulant for working *up* to a market. Henry Ford proved that long ago. A huge potential market existed for cars when Ford began his large-scale production : Ford being a man of vision knew that a car at a low price would bring that market to life. He fixed the price, then organised production to secure the enormous demand that he knew a low price would create. The size of this demand enabled him to reduce production costs, for when you are dealing with millions of cars the saving of one cent on a major or minor item among the thousands of parts that make up a complete car, and the saving of seconds on the time of an operation in the production sequence become astronomical totals in dollars, spread over a year.[1] Ford never cut down quality; he cut down waste in motion, time and material. He always sought out materials over which he could have complete control—materials which would not disturb the pace or sequence of production by incalculable behaviour. For example, twenty years ago, Ford gave up using wood for steering wheels because it was wasteful " and no wood-working operation can be carried through with absolute precision."[2] He used straw waste and, with a rubber base, sulphur, silica and other ingredients, produced a plastic material called Fordite, which could be worked without fear of disturbing fluctuations: also he saved about half the cost of the wood formerly employed.

Ford had to guess the size of his potential market. We *know* the size of ours: we know that we are nearly four million houses short. We know that mass production works well in the provision, not only of motor cars, but of thousands of other commodities. Unfortunately, their ignorance or deliberate disregard of the economic laws of mass production make many political and social reformers not only ridiculous, but obstructive. As they persist in trying to examine the housing problem in a controversial mood, obsessed by the rival claims of private and state enterprise, they forget the implications of the word *enterprise;* also they ignore technical facts, which, if fully apprehended, would suggest new economic solutions. It is only common sense to ask how much will the factory-built house cost; but although a specific answer in pounds, shillings and pence is not yet available, costs may be examined and some indication given of the extent of savings in labour and materials secured by factory production. Until we have long experience of the large-scale production of such houses, it is impossible to estimate a price. Up to the end of 1944 experimental models only had been produced, comparable to the early cars built by Henry Ford before he perfected the famous Model T, which ultimately brought cheap, efficient cars to people everywhere.

Since partial prefabrication has existed for many years in traditional building methods, an analysis of the cost of a typical council house of 1939, is instructive and significant. The costs are estimated broadly as 44% site labour and overheads, 25% labour in prefabricating materials, 31% raw materials. These percentages are distributed among the various trades as follows:—

---

[1] The Ford Model T contained about 5,000 parts, including screws and nuts. *My Life and Work*, by Henry Ford, Chapter V.

[2] *To-Day and To-Morrow*, by Henry Ford, Chapter VI. (First published in the U.S.A., 1926.)

	Labour and Materials	Labour	Materials
Excavator and concretor	5	2.5	2.5
Brickwork	35	13.5	21.5
Carpenter and joiner	22	7	15
Roofing	5	1.5	3.5
Plastering	7	4.5	2.5
Plumbing	8	2	6
Lighting	4	2.5	1.5
Glazing / Painting	3	2	1
Paths, drains and fences	5	2.5	2.5
Overheads	6	6	—
	100	44.0	56.0

The aim of the factory-built house is to reduce the site labour percentage to a minimum and by standardisation and mass-production to reduce to a minimum cost the prefabrication of the raw materials. The appropriate selection of materials for that purpose becomes an important cost factor.

In the figures just given the items " Brickwork " and " Carpenter and joiner " constitute by far the heaviest total percentages and would seem to offer the most important opportunities for prefabrication. As traditional site plumbing represents only 2% labour, the factory contribution here would appear to be almost negligible. But two considerations modify this conclusion : (1) *in situ* plumbing can be the source of a serious bottleneck in traditional building, and (2) the traditional rough-and-ready fixing of drainage and supply pipes on external and internal walls respectively had attained no satisfactory degree of efficiency up to 1939. Also, the amount of apparatus represented by the figures of percentages was the very minimum.

The standard of plumbing and equipment in the Portal house is far in excess of anything hitherto installed in any council houses in this country. There may be two good reasons for this : firstly, that in compensation for sub-standard accommodation and planning, equipment of a higher standard than that normally provided should be offered, and, secondly, which concerns the subject of this book, that since factory production is being employed, far greater amenities become available than could ever be contemplated by traditional building methods.

Although it is not possible to give actual figures or estimates for the mass production of the various prototypes of factory-made houses now being evolved in Britain, the fact that many industries have devoted serious attention to such production implies that reasonable prices can be envisaged. The costs of traditional building having risen 153% from 1916 to 1944—according to calculations made by Mr. H. J. Venning, F.S.I.—the level of such costs in the immediate post-war years is unpredictable. In traditional building, 500 houses cost less than say 5 or 50; but little reduction in price is possible by building larger quantities than 500. With mass production methods for the factory-made house, quantity can decrease price to a far greater extent. It is possible to visualise too that such decrease in price for the component parts could go to the provision of larger floor space and greater amenities.

As the working-class house has received from time to time more amenities, the relative value of a square foot of floor area has decreased in cost. A survey of housing costs made on pre-1939 price levels by the Ministry of Health showed that a 14% increase in floor area increased the cost by 6%.[1] With the factory-made house an increase of floor area, by using extra mass produced parts, would probably show considerably less increase in cost. Investigations of the cost of the factory-made house should not be primarily concerned with initial outlay on materials; they should be concerned largely with time. Time saved means money saved. Reduction in man-hours and machine-hours means reduced costs. For example: in most types of factory-built houses, the heavy central brick core which carries the flues and forms the chimney breasts becomes unnecessary. The flues may be completely independent of the structure, and can convey warmth to the rooms of an upper storey from the heating appliance on the ground floor. A more efficient coal-burning stove or open fire becomes possible; it is fitted into a simple aperture, joined up to the flues. The wrong way to examine the cost of this is to say: " But the stove or coal-burning appliance is of a much higher standard, and it costs several pounds more than the old-fashioned type." The right way is to consider the savings in man-hours secured by the omission of that mass of brickwork for flues and chimney breasts: the few extra pounds required for better coal-burning appliances are more than offset by lower labour costs. Reduction in man-hours and in the volume and character of the building materials needed are the major factors which establish the factory-built house as an economic

---

[1] Quoted from " Science and Housing," a paper read before the Royal Institute of British Architects on June 13, 1944, by Mr. A. M. Chitty, F.R.I.B.A.

HOW MUCH WILL IT COST?

proposition; and this reduction is secured not only by the technique of industrial production and by the use of materials which are economical to manufacture and can be handled with precision—thus making for easy erection—but by *the initial operation of design*.

The nature and extent of the savings in time and money secured by factory-made houses are suggested by the research work that has been carried out by the New York John B. Pierce Foundation. This became a corporate body in 1929, and was founded for the purpose of receiving and administering funds for research in the general fields of heating, ventilating and sanitation for the advancement of hygiene and for increasing standards of comfort for human beings and their habitations. Prefabrication for domestic building in the U.S.A. has already derived great benefit from the work of the Foundation. It is referred to here because its statements may be accepted without question or doubt, and the cost figures for the completely equipped houses it has already sponsored are of practical interest.

The first test model of a house was made by the Foundation in 1939, after a long period of research and experiment. A second slightly improved model then followed, and finally a village of twelve newly designed units was constructed and the houses were rented to families. From this experience it was found that a single-storey dwelling (floor area 24' by 32') comprising living room, three bedrooms, bathroom, and kitchen with heating plant, electric cooker, refrigerator, water heater and a considerable amount of built-in furniture (value $130) could be provided for $2,079. The same sized floor area with one bedroom less would be $200 cheaper. Such a house is of timber construction but it is not designed for short life, since it is intended for purchase on a 25-year mortgage. Little money is spent on foundations, which consist of ten concrete piers, 3' deep in the ground, poured *in situ*. *This work involves* 350 *man-hours*. The construction itself consists of scientifically designed timber framing with $\frac{5}{8}''$ resin-bonded plywood structurally embodied with it. An interlining of 1" glass wool blanket is used for insulating the walls, and 2" thick blanket over the ceiling. The interior linings are of $\frac{3}{8}''$ plywood squares pre-painted in the factory and bevel-jointed. The low pitched roof construction is sheathed with $\frac{1}{2}''$ plywood covered with building paper and asphalt shingles rapidly stapled in place with self-feeding hammers.

These houses can be decorated and finished with heating, plumbing, electric wiring and built-in furniture *complete within ten days of erection* beginning.

Elevation and plan of the Pierce Foundation experimental house. (See page 95)

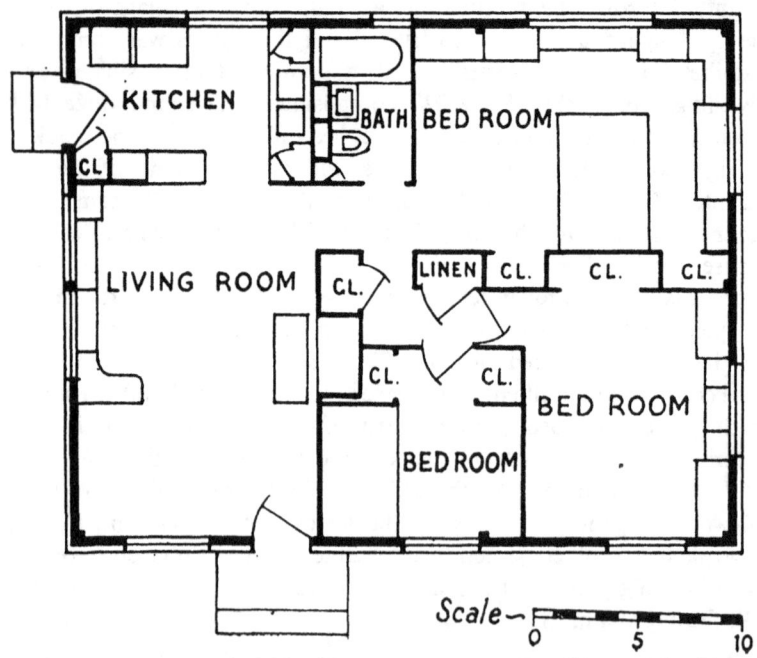

Alternative schemes have been worked out for heating by coal, or wood, oil or gas. The electric installation is of the new plastic-covered conduit which is described in Chapter VIII. The external appearance of the house follows the traditional American colonial style ; the interior has a high standard of finish. A further scheme of twenty-two low-cost houses was built in New Jersey in 1940 to gain additional experience ; after a year's occupation most favourable reports were received from the tenants.

Some instructive figures relating to prefabricated dwellings imported from Sweden were given in the *Architects' Journal*, early in 1944, by C. Sjostrom, A.R.I.B.A.[1] These figures were at the 1938 level and were for a prefabricated semi-detached bungalow, designed to meet Scottish standards, for erection in Scotland, with a floor area of 663 square feet. The total was £358, itemised as follows :—

	£	s.	d.
Excavations, foundations, base wall, chimney stacks, and party wall	51	0	0
Timber sections, windows, doors, partitions, joists, boarding, ironmongery, transport and customs duties	150	0	0
Erection of timber structure	30	0	0
Expert supervision	7	5	0
Wallboard and fixing	21	10	0
Cedar shingles to roof	19	10	0
Sundries	12	15	0
Plumber and sanitary fittings	47	0	0
Painter	11	3	0
Glazier	2	2	0
Electrician	5	15	0
	£358	0	0

This represents 10s. 8d. per square foot floor area.

In considering the justification for emergency dwellings, which are uneconomic with their short term life, existing housing conditions must be considered. In the County of London, the 1931 census reports by the valuer to the Housing and Public Health Committee, 1935, published by the L.C.C., reveals that nearly two-thirds of the private families in the County (752,705 families out of 1,190,030) were accommodated in shared dwellings, where in many cases the provision of bathrooms, w.c.'s, kitchens and water points in the buildings were inadequate for the use of even a single family. The shared dwellings over the whole county averaged 2.56 families in each. 167,130

---
[1] February 3, 1944.

dwellings had four or more families occupying them, that is, half-a-million families sharing such services as existed. In this census it was shown that 13.13% of the private family population were living more than two to a room and 47,343 persons were living more than four persons to a room. There was one instance of eleven persons occupying a single room.

In Battersea it has been estimated that 60 to 70% of the w.c.'s in the Borough are used by more than one family.[1] In Shoreditch four hundred school children, selected at random, revealed that only 25% lived in houses or flats with separate w.c.'s, for each family, and in 50% of the cases the closet was shared by a number of families ranging from two to seven.[2] The survey in respect of water supply, heating and bathing facilities showed that in 367 cases one in three had no indoor water supply, and that 30% of the families had to carry water up two or three storeys. In 402 cases nearly half the families had no other means of water-heating than by kettle, and out of a similar number only 14% had a proper bath. Of the remainder, two-thirds had not even the facilities of a tub. Such conditions are not restricted to London : provincial towns and cities and even picturesque country slums can produce shameful statistics of over-crowding, obsolescence and low standards of housing. (Since these statistics were published, the housing conditions in London and elsewhere have still further deteriorated through war damage.) But perhaps the most terrible indictment of our incapacity or unwillingness to use the immense resources of industry for turning out good houses with the same ease and rapidity with which we turn out good cars, is the survey of the conditions of 1,250 married working women, by Margery Spring-Rice, based on information collected by the Women's Health Enquiry Committee, and published by Penguin Books under the title of *Working-Class Wives*.[3]

During 1944 the possibilities of mass production were at least officially recognised ; but recognition was accompanied by a complete misconception of the permanent nature of the benefits it could confer upon housing. It was regarded as a useful " dodge " for meeting an emergency : its true significance escaped the politicians and officials, though perhaps the placating of vested interests, both in the established building industry and the building trade unions, demanded a little judicious blindness. But it was to meet an emergency only that

---

[1] *Our Towns—a Close-up.* (Oxford University Press, 1943.)
[2] *Growing up in Shoreditch.* (Shoreditch Housing Association, Ltd., Toynbee Hall, 1938.)
[3] First published in 1939.

Lord Portal, on February 8th, 1944, announced that the Government proposed to provide, by mass production, large quantities of a short-term house constructed mainly of light steel. A major consideration in launching such a scheme was that the houses, being sub-standard in construction and accommodation, a short-term life was assured and that there was little danger of their remaining in use for years either as slums or eyesores. But having launched such a production programme, the Government soon discovered the need for including other types of mass produced construction for the same objective. Some of these systems of prefabrication will provide dwellings with a far longer life than the emergency house was intended to enjoy.

That such dwellings need not be ugly was proved when the United States, faced with the necessity for emergency houses, particularly in the T.V.A. scheme, achieved some most agreeable results. Some of these houses were designed for a three-year life only, but are now considered capable of a ten-year life, with reasonable attention.

The immense devastation in Russia apparently stimulated titanic efforts in providing housing, even during the war period. To achieve this, encouragement was given to anyone to lend a hand ; credits being granted by the Government to assist the work, while local trade unions helped with technical advice, tools and transport. This example of Russian social and economic realism may be compared with a statement by Mr. V. Sulston regarding the British trade union attitude to prefabrication when, as London Secretary of the N.F.B.T.C., he was quoted as saying : " There will be the biggest row in history if the job [of putting together on the sites] is not confined to building trade operatives."[1] It was reported that during the first nine months of 1943, 25,000 prefabricated houses were built in Russia. Many factories were also established for producing prefabricated dwellings. While factory production played a considerable part in saving transport, local building materials were used as much as possible.

In September, 1944, the Parliamentary Secretary of the Ministry of Works, in a statement to the Press, said that of a pre-war house, total cost £461, materials represented £313.[2] These materials to-day cost £501. The rise in prices is made up as follows :—

Rise in Timber .. 160%
Rise in Bricks .. 45%   (since this statement was published it has risen to 60%).
Rise in Cement .. 44%

---

[1] *Architect and Building News*, April 7, 1944.
[2] *Architect and Building News*, October 6, 1944.

These rises in material costs, amounting to just over 60% are due mainly to increased cost of transport fuel and the heavy overheads of war risk insurance.

The Ministry of Works estimate for a three-bedroomed house (862 square feet) such as the specimen built at Northolt in the autumn of 1944, is £759, based on building 500 at a time on a site. For this specimen house, the site man-hours were approximately 2,100. The construction was traditional with 11" brick cavity walls.

The cost of traditional building has mounted. The percentage rises in labour costs over those obtaining in September, 1939, has been given by Mr. H. J. Venning, F.S.I., as follows :—

(a) National increases on hourly rates, based on cost of living   30%
(b) Guaranteed week, holidays with pay, increased insurances, site welfare, tool money   5%
(c) Cost of importing men from higher rated districts, designated labour, travelling money and time, free transport, subsistence, etc.   40%
(d) Overtime and exceptional margins on jobs   7%
(e) Drop in output   15%

                                                                                       97%

He suggests that the increased percentage on items (a) and (b) may possibly persist and may constitute a total 35% rise in future costs of labour.[1]

From these two estimates of rises in the cost of labour and use, and in the cost of materials, traditional building in the future may cost between 40% and 60% more than in 1939.

In the face of these inexorable facts, what chance is there of being able to build £300, £400 or £500 houses by traditional methods ? Only if they are the merest shells, with scamped finish and poor equipment. Even then it is doubtful whether in ten years we should even begin to meet the need for houses, apart from denying to a people who deserve them the easily-run, well-equipped, warm, insulated, pleasant homes that could pour out of the factories in their thousands, every day, if we had the will and the wit to use the means that exist. A new building industry is needed, composed of enterprising manufacturers working in collaboration, who would approach this huge, known and promising market as Henry Ford approached the market which he believed in, though he could not measure it save by his knowledge of human nature. A car manufacturer knowing that a

---

[1] *Architect and Building News*, February 25, 1944.

market for four million cars existed, at home, could put a remarkably efficient cheap car on the market and reasonably quickly. Now a car is a rather complicated little house on wheels, and it has far more mechanism than any dwelling house carries; moreover, it is not a static structure, and it has to stand up to all the stresses and strains that mobility implies. It has to resist all the weather conditions that a house normally resists, but for a car those conditions are often intensified, because it is being driven *through* rain and snow and fierce gales of wind. Yet this mobile structure, with all its mechanism, and its comfortably upholstered furniture, its lighting and complete equipment, costs very little. In 1939 the Morris Eight two-door fixed head saloon cost £128; the Ford Eight standard saloon, £115; the Ford V-8, 30 H.P. saloon de luxe, £280; the Vauxhall 12, standard saloon, £189; the Vauxhall 14, £230; and the Rover 10 H.P. saloon, £275. These prices are instructive. They suggest what good design and industrial production can achieve. If the market is represented by a demand running into millions, need a factory-built house cost much more than a Morris or a Ford car? An industrialist with vision—a Nuffield or a Henry Ford—might even ask, need it cost as much?

A group of industrialists, experienced in fabricating metal, plastics and other precision materials, could produce factory-made houses for assembly, free on site, both quickly and cheaply if they had good designs and were able to book orders in advance from local authorities who had housing schemes in hand. If they could budget for only half a million of such houses, they might bring the price down to £X a house, including accommodation and equipment at least as good if not better than that used in the Portal house, and provide a dwelling with a longer and better life than that anticipated for the short-term prefabricated house. The cost of this house must for the moment stand at £X; but any industrialist with vision and experience could turn that symbol into a figure that would absolve the State, and the taxpayer, from the burden of subsidies. Local authorities, government officials, hard-worked statesmen, and—most important of all, and *for whose safety, welfare and freedom the State exists*—the public, should study the cost of making motor cars in thousands, and ask, and continue to ask, why industry should not give to home-makers what it has given to motorists. If we created this new housing industry, not only could our own needs be satisfied, but a new, growing field of employment would be prosperously developed, and, sustained by a flourishing home market, a great export trade in factory-made houses might arise, for we are not the only country that is suffering from a severe housing shortage.

CHAPTER 7

Will it be Comfortable?

BEFORE 1939 the great majority of houses in Britain were far below our present-day standards of insulation, heating and plumbing. The housewife had gained little from the progress and improvement of industrial production, except in superficial equipment. That equipment was certainly spectacular, and because it reduced the volume of arduous housework it diverted critical attention from the fact that in substance and structure, in design and general arrangement, the house itself often created a permanent background of inconvenience and small, pinprick annoyances. A sewing machine, a gas or electric cooker, a vacuum cleaner, an electric iron, a refrigerator, a kitchen cabinet, might help to smooth " the daily round of irritating concerns and duties " ; but even those industrial products were denied to thousands of homes, and it was unfortunately and quite absurdly true that most of the small houses in this country differed from their eighteenth century equivalents only in the possession of water closets, a good water supply, and convenient artificial lighting.

During the last hundred and fifty years nearly every department of life except housing has derived revolutionary benefits, directly or indirectly, from the growth of industry. From the end of the Georgian period to the present day a spate of inventions has grown in strength and volume until it has washed away nearly all the established methods of manufacturing, locomotion, illumination and communication, thus giving to the commercial machine age in which we live a character unique in history. No former civilisation could command the comforts and labour-saving devices that are now available for everyone. The Romans depended on slaves, Mediæval Christendom on serfs, the Georgians and Victorians on underpaid domestic servants : the results in wilfully wasted human lives were almost identical. Machines should relieve us all from the personal drudgery that in previous ages was avoided only by the privileged. But we seldom use all the advantages that are available.

In many ways the so-called modern house is as remote from modern standards of equipment and performance as the clumsy seventeenth century stage wagon is remote from the long-distance motor coach, or the *Golden Hind* from the *Queen Elizabeth*. In house building we have, so far, rejected the principles of large-scale industrial production, and although, as we have already said, many items of individual house equipment are mass-produced, their production is not related to a system for manufacturing the whole house in prefabricated parts. We know so much about the making of comfortable and efficient homes to-day, and up to 1939 we used so little of our knowledge, that surely only ignorance of the economic and technical facts could account for such consistent rejection of the higher standards of living which are available for the British people. By rejecting these standards now we are defrauding future generations; we are condemning the children of the second half of the twentieth century to lives in mean surroundings, and we are throwing away the results of national and industrial research.

It is not only abroad that research is conducted into new methods of constructing and equipping houses, though critics who love to say, " my country, always wrong," ignore the enormous volume of research that is carried out here, and which is often partly or wholly financed by British industry. It has been the fashion for highbrow writers to assume that industry is run by greedy, incompetent and anti-social sub-men; a fashion which reflects what Mr. Winston Churchill has described as " the mood of unwarrantable self-abasement into which we have been cast by a powerful section of our own intellectuals . . ."[1] It denotes a calculated falsification of the facts of industrial and commercial life, as obtuse as the faith of those who believe that the earth is flat, though lacking the childish innocence which flouts geographical realities. The Department of Scientific and Industrial Research, the National Physical Laboratory, and the Building Research Station, plan and carry out vast programmes of research, in collaboration with various industries, and test new materials and methods for using them. There are many other research organisations, connected with specific industries, such as the British Coal Utilisation Research Association, sponsored jointly by the coal industry and the manufacturers of coal-burning appliances; the Forest Products Research Laboratory, supported by the State and to some extent by the timber industry; the British Cast Iron Research Association, started and financed by the ironfounding industry—these are only a few examples of active research bodies. There are many others.

From these, as well as from the research laboratories of individual firms, a stream of technical information flows out which should refresh

---

[1] Mr. Churchill made that statement in a speech to the Royal Society of St. George April 23, 1933.

the inventiveness and efficiency of the building industry and increase the comfort and welfare of the people. For example, it is only within the last quarter of a century that the contribution structural materials can make to the warmth and comfort of a dwelling has been the subject of scientific study. From this study a great body of facts about thermal insulation has become available, leading to one of the most important single developments in building technique that have occurred in our time. A leakage of water is easily seen and stopped; a leakage of heat rarely receives attention, though it could be corrected by the proper insulation of buildings and appliances and by designing appliances to operate economically.

" Given good insulating walls, the problem of heating houses entirely disappears," writes Professor J. D. Bernal. " Indeed, even in winter, the heat generated by the inhabitants of the houses would require some method of cooling to get rid of it. To secure this degree of self-sufficiency, however, it would be necessary to devise a rational ventilation system which did not, as at present, take in air cold and send it out hot, but arranged for the outgoing hot air to warm the incoming cold air in winter time and vice versa in summer."[1] He also points out that " already reversible heat engines which would pump heat into a house in winter and out in summer have been run on one-third to one-fifth of the cost of direct heating methods."

To achieve good insulation the house designer has to consider many factors concerning material, construction and cost. For instance, an air space whether $\frac{1}{2}''$ or 4" wide is only equal to $\frac{1}{2}''$ to $\frac{1}{3}''$ thickness of cork. A 6" thick wall of foam slag and a 27" thick brick wall have equal heat transmittance values. The required output of a heat appliance, again, is dependent upon how quickly the inner lining material of a room will warm up, and a great deal more heat will be needed to warm up the 27" brick wall than the 6" foam slag wall. It takes, in fact, ten or more times as long to heat up a brick-walled room from 30° to 60° as to heat one with an inner lining 1" to 2" thick of grade " A " insulation.[2]

There is also the problem of water absorption in outer walls. A brick wall 9" thick may absorb 20-lb. of water per square foot of surface. If 5-lb. of this is absorbed internally by conversion into vapour, some 5000 B.Th.U. per square foot will be used up in this conversion. Such use of fuel to absorb moisture from walls is wasteful, since a damp-proof inner lining would render it unnecessary.

[1] *The Social Function of Science*, by J. D. Bernal, F.R.S. Chapter XIV, p. 351. (Routledge, 1942.)
[2] *Heat Insulation in Domestic Buildings*, by Colonel S. F. Newcombe. Journal of the R.I.B.A., August, 1944, pp. 253-260.

Surfaces that heat quickly are of particular benefit in the English climate where sudden changes of temperature, up or down, are frequent and demand intermittent heat. Therefore, the most efficient forms of construction are those which consist of a hollow space with an inner lining. The Burt Committee proposes maximum values of heat transmittance for walls of houses.[1] The adoption of such standards could save valuable fuel, but the importance of the recommendation is its recognition of the need for a scientific basis for this aspect of housing. The various grades and capacities of insulating materials afford considerable choice for the factory-made house, which would enable them to attain standards of efficiency in insulation that are seldom reached by traditional building methods.[2]

The traditional brick walls of a house are generally stronger than structurally necessary, but do not give efficient insulation or protection from moisture. The pitched roof on the traditional house (unless enclosing an attic room) merely serves as an umbrella protection from the weather. It should have no insulation wasted on it; the insulation should be in the flat ceiling below. A weather-tight flat roof is a logical answer for many types of factory made houses, and has no disadvantages, apart from offending the taste of people who forget that many of the eighteenth century houses they admire had flat roofs, behind the coping or decorative stone balustrade that terminated their exterior walls. Good design and familiarity would in time mitigate objections to a very old and practical method of roofing a dwelling.

For general guidance on room temperatures in our climate and methods of securing them that great authority, Professor Alfred Egerton, may be cited.[3] He recommends a temperature of 65° F. for English people sitting at rest in their indoor clothes, though a house should not be kept at this temperature throughout. It would for instance be too high in rooms used for active work, such as the kitchen. The aim should be to avoid such extremes as overheated kitchens, chilly living rooms in the early morning, and icy bedrooms. In improving the standard of comfort it is important to avoid getting people accustomed to overheating. For the living room he suggests a temperature of 65° F. for $9\frac{1}{2}$ hours of the day spread over three periods, and of 55° F. for 6 hours in two periods. The hall should be at 55° F. for $15\frac{1}{2}$ hours, and the bathroom at 60° F. for half an hour.

---

[1] *Post-War Building Study*, No. 1, para. 183. (H.M.S.O., 1944.)
[2] Useful gradings of insulating materials and their performances were given in *Heating, Piping and Air Conditioning*, June, 1942, and are fully quoted in Colonel S. F. Newcombe's article in the Journal of the R.I.B.A. See footnote on page 104
[3] Paper read at the Institution of Mechanical Engineers, September 20, 1944.

Houses in the North may need 20% more heat to secure this result than those in the South, but South rooms may require 10% to 15% less heat owing to their aspect. A background of heating, equivalent to 50° F. should be planned, the extra temperature needed at certain times of the day being obtained from easily controlled appliances.

Professor Egerton places in order of efficiency, stoves and methods of central heating and flueless gas fires as giving fairly high results, closable fires, small coke fires, and gas fires come next, and open fires and electricity in the lowest category. He considers the open fire to be not so much a room warmer as a ventilator, and as such it is a luxury. Suitably controlled central heating is recommended as the most rational method of producing background warmth, and it needs little additional heat from other sources for attaining an adequate temperature. There is apparently much scope for development of the small, efficient, convenient and easily installed central heating boiler, needing attention only once in twelve hours. Such an apparatus should, in Professor Egerton's opinion, be able to supply hot water also for the household. It must be recognised that the open fire is a very old institution; it may, as Professor Egerton suggests, be a luxurious ventilator; but there is a sentimental attachment to it that has survived centuries of reasoned protest against its wastefulness and dirt. In the mid-seventeenth century " Inventive gentlemen, like Sir John Winter, might attempt to popularise a form of coke, so that coals could ' make a cleare pleasant chamber fire, depriv'd of their sulphur and arsenic malignity,' but the glowing comfort of a coal fire was too pleasant for most people to bother about the smoke it made. In farmhouse and cottage the huge fireplace was really an alcove in the living room, with seats or high-backed settles on either side, but in those fireplaces wood was burned, and the chimney mouth would be used for curing bacon in the woodsmoke. The ' chimney corner ' was the most sociable, comfortable place in the small house.

" Hearth and home have always been closely connected in the minds of Englishmen, and even the Puritans did not attempt to disturb or forbid the lighting and enjoyment of fires."[1]

Many of the practical suggestions for improved heating methods in buildings came from gardeners. The pioneer was Sir Hugh Plat. As early as 1653 he used a boiler on a kitchen stove to convey steam heat through pipes to growing plants, thus rendering them independent of outside temperature changes. In 1745 Sir William Cook improved on this method and published a diagram for heating all the

---

[1] *The Englishman's Castle*, by John Gloag. (Eyre & Spottiswoode, 1944.) Chapter IX, p. 89.

rooms in a house from the kitchen fire. In 1784 James Wyatt used a very elementary type of steam radiator to warm his study, and in the middle of the next century Sir Joseph Paxton produced a design for a crystal sanatorium which embodied not only an installation of filtered and heated air, but had the additional benefit of extra oxygen from growing plants.[1] This was the birth of modern air conditioning.

There are two major considerations concerning heating appliances. The first is to eliminate waste in coal consumption; the second is to reduce atmospheric pollution by smoke in built-up areas. Mr. R. Fitzmaurice of the Building Research Station has stated that "the value of coal consumed annually in heating the houses of Great Britain would suffice at 1938 prices to build 200,000 to 300,000 houses ... conservation of heat depends upon the efficiency of heating appliances and on the insulation of buildings."[2] Incidentally, if each of the 12,000,000 houses in the country saved a ton of coal a year, it would be equivalent to a capital saving of £400,000.

The pollution of the atmosphere by coal fires has been worrying legislators and reformers and technical specialists for over six hundred years. Edward I in 1306 forbade by proclamation " the burning of Sea coale in London and the Suburbs, to avoid the sulferous smoke and savour of that firing . . ."[3] Three hundred and fifty-five years later John Evelyn wrote his famous pamphlet, *Fumifugium*, which he dedicated to Charles II, condemning the use of sea coal in London and bringing to the support of his case an array of scientific and reasonable arguments, which are still valid. Both Edward I's proclamation and Evelyn's pamphlet were directed chiefly against the industrial use of coal in the city. But to-day nearly half the atmospheric pollution in Great Britain is caused by domestic coal fires. Every year they pour into the air above the chimney pots about 1,000,000 tons of smoke and huge quantities of sulphur dioxide. It has been calculated that this output of sulphur dioxide represents as much corrosive sulphuric acid released in the form of smoke as that produced annually by the chemical industry of Great Britain. This emission of sulphur dioxide can be reduced by cleansing the coal before burning. Serious attention is now being given both to the uneconomic use of coal and to atmospheric pollution by smoke. Invaluable research has been done during the war both by the Department of Scientific and Industrial Research and the British Coal Utilisation Research Association.

---

[1] *Illustrated London News*, July 5, 1851.
[2] Statement made at the Cantor lecture, Royal Society of Arts, in 1943.
[3] Quoted from Howe's second edition of Stow's *Annales*, published in 1631.

The measurement of smoke emission was first achieved only a few years before 1939. Several revolutionary designs for grates have been evolved which, while maintaining the open fire, will reduce smoke considerably—possibly as much as 75% of normal emission—and will remain alight for many hours without attention; these appliances also secure a substantial saving in fuel. It has also been realised that considerable heat might be distributed by convection (that is as warm air) with suitably designed appliances. The scientific distribution of this heat to other parts of a house make only one solid fuel appliance necessary. The provision of extra flues for conveying the heat is apt to be costly and difficult with traditional construction, though admirably suited to the flexible construction of the factory-made house, whose light frame can be arranged to carry independent flues and ducts wherever required. Few members of the public had seen any demonstration of convective heating before they inspected the sample Portal house. The construction of this house was so well suited to such heating that it was installed, despite the fact that in Britain it was revolutionary, though long-established in practice on the Continent. The correct insulation of the appliances also secures savings in fuel consumption. Tests conducted at Illinois University, U.S.A., have shown that 6.5% saving of fuel is possible by placing reflective insulation behind radiators, and a 10% saving of fuel possible by using a 2" thickness of suitable lagging behind an open grate.

That the factory-made house could be warm, weatherproof and comfortable is indisputable; and in the living room improved appliances could give visible heat with efficiency and economy, thus allowing traditional notions of comfort to be maintained. Our national liking for the fireside may be sentimental; but it is none the worse for that, and it is a liking that should be respected. Factory-made houses should be designed not only to use modern materials and industrial technique in the most effective way, but to serve the known and established taste of the people who will make their homes in them; and those people like comfort, privacy and the intimacy and cosiness of their own fireside.

CHAPTER 8

Will it be Easy to Run?

BUCKMINSTER FULLER'S Dymaxion house has been mentioned in Chapter 3, and some of its characteristics may be appropriately considered now, for it is designed to be as easy to run as a motor car, as self-contained and independent as a mediæval manor house, and it could be factory-produced at high speed by the million if the foresight and courage which engender enterprise were available. Its designer is an American engineer, and in 1929 he produced this unusual, original and startling solution to the problem of making a house with the fullest use of the most appropriate contemporary materials and scientific achievements. The name *Dymaxion* is compounded of *dynamics* and *maximum service.* Buckminster Fuller was not hampered by any preconceived ideas about architecture. People had to live, they needed shelter, and an abundant and beautiful diversity of materials, methods and systems existed, which had arisen from a century of continuous and expanding scientific and industrial research ; so to the mind of an inventive engineer the problem was relatively simple, and he set about selecting what would give the best service in the least troublesome way to the inhabitants of the house.

The Dymaxion house is a piece of architectural idealism, as much in advance of its time as Jonathan Hull's steamboat, which he patented in 1736 ; the steam carriages of Oliver Evans, for which he obtained an exclusive right from the state of Maryland, to make and run, in 1787 ; or the flying machines visualised by Leonardo da Vinci—but with this difference : the early ideas for the steamboat and the steam carriage had no great accumulation of tried and tested mechanism to make their launching on the water or running on the roads an immediate and practical possibility, and Leonardo's vision had to await the advent of the internal combustion engine for fulfilment. For putting projects like the Dymaxion house into industrial production *we only await the will.* " All things are ready, if our minds be so." Meanwhile the house that makes full use of the scientific facilities of the twentieth century remains in the idea stage, a smouldering incitement to revolutionary design.

This highly mechanised house has no bath : an air pressure hose squirts 90% air, 10% water in a cleansing stream—no soap is needed. There are no taps or sinks or water closets—toilets are of the dry type used in passenger aeroplanes. A machine converts sewage into methane gas to provide light and power. All machinery fits into a central aluminium mast. It is air conditioned : bedclothes are unnecessary. Beds are pneumatic and cupboards have revolving shelves. Walls are transparent but windowless. All doors open by waving the hand before photo-electric cells. Cooking is done by vacuum Magda units. Dish-washing and laundry are mechanical. The house can be planted anywhere regardless of sewage disposal, gas or electric supply. There is a shield the whole height of the building which swings with the wind. The mast is lashed by guy wires. A cone on the housetop admits air. The shape of the house is streamlined, since a square one would pile up pressure on the windward side and the vacuum thus created on the leeward side would suck out heat from the building. (See page 71).

It has been claimed that the Dymaxion house could be assembled in under an hour and that all the materials for it could be transported on one trailer. That claim was made in an article, published by the American magazine, *Shelter*, which discussed the production of Dymaxion houses for workers in the U.S.S.R.[1] In a concluding gust of jargon the article discloses the fact of very short life. " This shelter," we learn, " is to abet establishment of the ' heavy industry's ' plants, with a specific longevity of five years ; of integrally self-sustaining mechanics, and high mobility of placement." Which means, presumably, that it would be useful in housing workers in heavy industries, that it lasts for five years, is mechanically self-contained and can be sent almost anywhere for erection. The enthusiasm of many writers and critics for the Dymaxion house often curdled into almost unreadable descriptions, which bored ordinary practical people almost as much as the external appearance of the house repelled them. Eventually, such novel forms and the services they provide may become generally acceptable ; but there is always a time-lag between invention and familiar use, before the original idea that was once startling, even shocking, is a respectable part of

---

[1]*Shelter*, November, 1932, Vol. II, No. 5. Raymond McGrath devotes an informative section of his book, *Twentieth Century Houses* to Buckminster Fuller's design. Sec. 40, pp. 113–117. (London, Faber & Faber, 1934.)

everyday life. " No, my dear," said the old lady ; " I have *never* been in a motor car. I travel by train, as Nature intended I should ! "

But whether we find it shocking or satisfactory, we have been slowly moving towards something like the Dymaxion house ideal for several generations. Although the house, to a home-loving people, is not and can never be regarded as " a machine for living in," it contains mechanism. These " machine parts " of the house have only come into general use as the result of many inventions. Plumbing, heating, cooking and lighting appliances have been introduced gradually during the last three hundred years. They are still thought of as individual items ; and, as such, they are seldom co-ordinated with the design of the house. They are put in, like the furniture, independently.

Only over a long period have the appliances which we now consider essential for the health and comfort of every home secured their place. For example, the water-closet was re-invented by Sir John Harington in 1596 ; it had been originally invented in the Minoan civilisation, perhaps as early as 2000 B.C., but Sir John's invention was, so far as English civilisation was concerned, entirely original.[1] The inventor was so delighted with his ingenuity that he wrote a rather light-hearted description of its mechanism and advantages, entitled *The Metamorphosis of Ajax*.[2] But for the next two hundred years water-closets were rare curiosities. The first valve water-closet was patented by Alexander Cummings in 1775, a type that was perfected by Joseph Bramah in 1778. But it was only during the nineteenth century that the W.C. was considered an essential item in house equipment. By the end of that century, practically every house in the built-up urban areas of this country had at least one installed. The introduction of baths took longer. In 1894 only 3% of houses had them : by 1936 this figure had risen to 60%. Bathrooms were " put on," and the backs of many terraces show such additions clinging like barnacles to the upper walls.

The " putting on " or " fitting in " of plumbing and services are incompatible with the basic idea of factory-made houses. A new approach to the problem of installing plumbing and services is essential ; and the practical results of such an approach have been demonstrated by the architectural department of the City of Coventry, which

---

[1] " Crete . . . supplies the first known example of a water closet, showing traces of a wooden seat and a flushing arrangement." *Sanitation Through the Ages*, by Desmond Eyles. *The Official Architect*, April, 1941. Vol. IV, No. 4, p. 160.

[2] Some extracts from Sir John Harington's description were published in the special Sanitation number of the *Architects' Journal*, March 4, 1937, Vol. 85, No. 2198, p. 390.

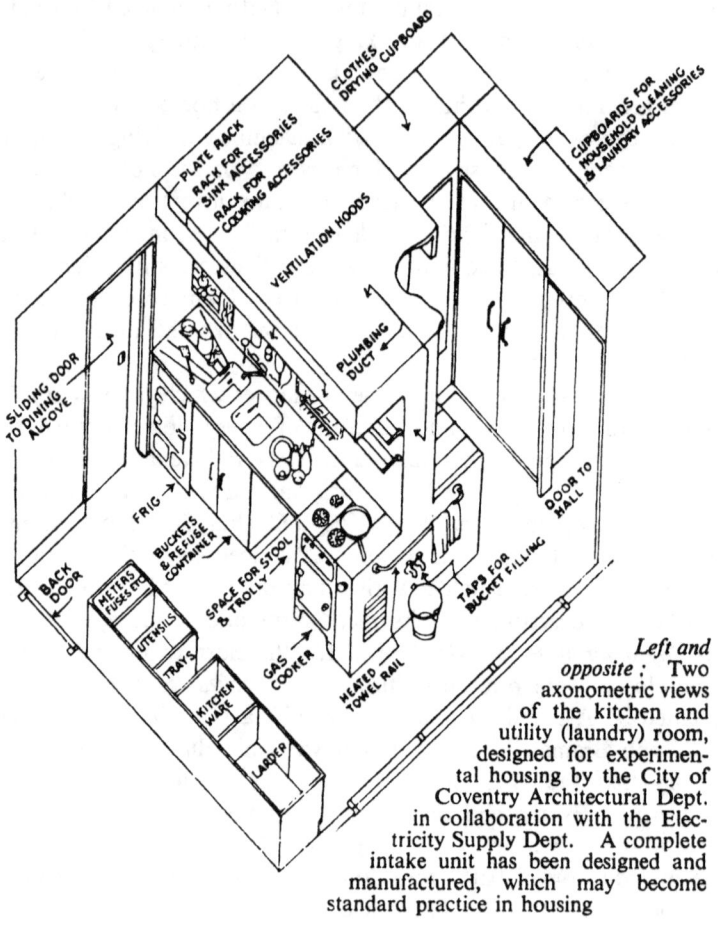

Left and opposite: Two axonometric views of the kitchen and utility (laundry) room, designed for experimental housing by the City of Coventry Architectural Dept. in collaboration with the Electricity Supply Dept. A complete intake unit has been designed and manufactured, which may become standard practice in housing

has, under the direction of Mr. Donald Gibson, F.R.I.B.A., evolved a central, self-contained plumbing and service unit. Here is a compact "machine part" for any dwelling, factory-made or built by traditional methods, as different from the bits-and-pieces technique of old-fashioned plumbing as a neat and tidy suitcase is different from an armful of parcels. For years the public have been used to festoons of pipes on the outside of buildings: they have accepted such excrescences as inevitable. When the pipes serving both drainage and supply freeze up and burst in winter, it's the winter that gets blamed. ("Unusually hard this year!") It is seldom realised that our plumbing is usually inefficient. Long runs of pipes within buildings are

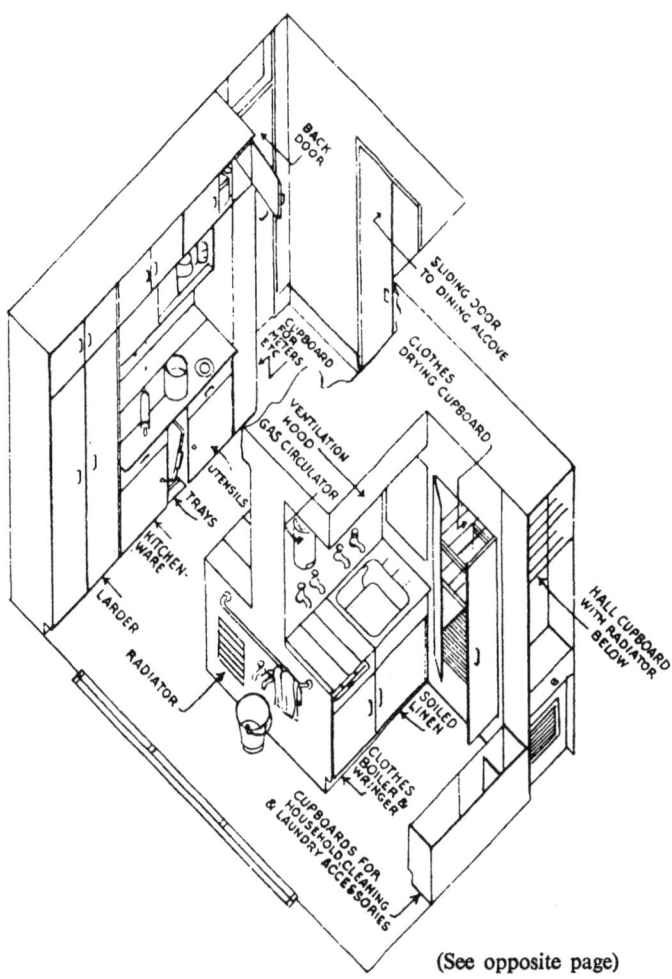

(See opposite page)

unsightly and dust-collecting if seen, inaccessible if hidden, and are the result of failing to plan the house round the central services.

Some years ago, when the one-pipe plumbing system was first introduced, a half-hearted attempt was made to group fittings round it, but local authorities were reluctant to accept such a system because of various rules and regulations about back-venting, which was obligatory, and put an unnecessary cost on the system. Some authorities did not even allow soil pipes to be placed within buildings. Recent experiments carried out by the Department of Scientific and Industrial Research and the Building Research Station have now proved that back-venting is unnecessary.

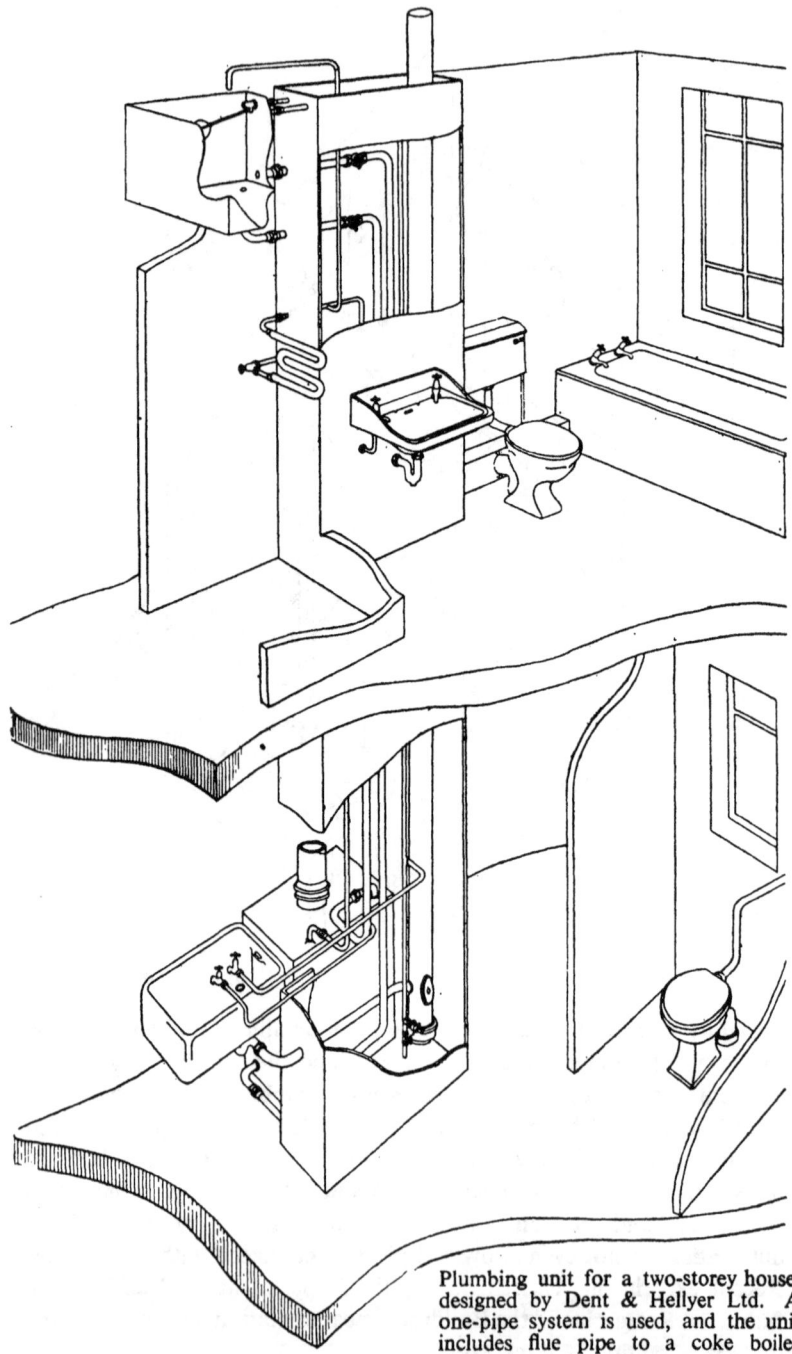

Plumbing unit for a two-storey house, designed by Dent & Hellyer Ltd. A one-pipe system is used, and the unit includes flue pipe to a coke boiler (*Reproduced by courtesy of Dent & Hellyer Ltd.*)

There is now no obstacle to prevent the use of a central plumbing and service duct, to accommodate all fittings with the shortest possible connections, becoming established as recognised practice. This entails planning around the plumbing and services but secures great economy and efficiency. Such a system is ideal for factory prefabrication : precision, proper examination of workmanship and testing can thus be easily achieved. The whole apparatus can be attached to a light skeleton frame, which can then be delivered complete or in sections to the building and only a few *in situ* connections would be necessary. All " making good " operations by plasterer, joiner and painter would be eliminated, and the plumber himself, instead of working in ramshackle sheds or on makeshift benches, could become a technician, enjoying ideal workshop conditions.

No house is easy to run without efficient plumbing. *But efficient plumbing is not enough.* There must be an unstinted and continuous supply of hot water. To provide this supply is the most practical way of giving every housewife more leisure, less arduous work, and a better chance of coping with the tasks that nag at every minute of her day. From the time she rises to get her husband's breakfast, wash the children, get their breakfasts and pack them off to school, until the time she finally washes up at night after supper she wants hot water. If people had boiling hot water, always available, ready in the tap, standards of personal hygiene would improve. Enormous numbers of people in this country don't wash either themselves or their clothes properly or regularly. An English crowd smells as no American crowd ever smells. There's better plumbing in North America, and higher standards of cleanliness. Eliza Doolittle's comment in Shaw's *Pygmalion*, after she had had a hot bath in a proper bathroom for the first time in her life, is an indictment, not of the habits of the poor, but of the totally unnecessary difficulties that beset their lives. She said : " I tell you, it's easy to be clean up here. Hot and cold water on tap, just as much as you like, there is . . . Now I know why ladies is so clean. Washing's a treat for them. Wish they saw what it is for the like of me ! "

Hot water, and the means of heat for producing it, may be obtained in a variety of ways to-day. There are the alternatives of solid fuel, oil, gas and electricity. The hot water supply may come from outside the individual dwelling—a method known as District Heating—or it may come from within, furnished by an independent appliance, installed either centrally or locally or as a subsidiary from a cooker or space heater.

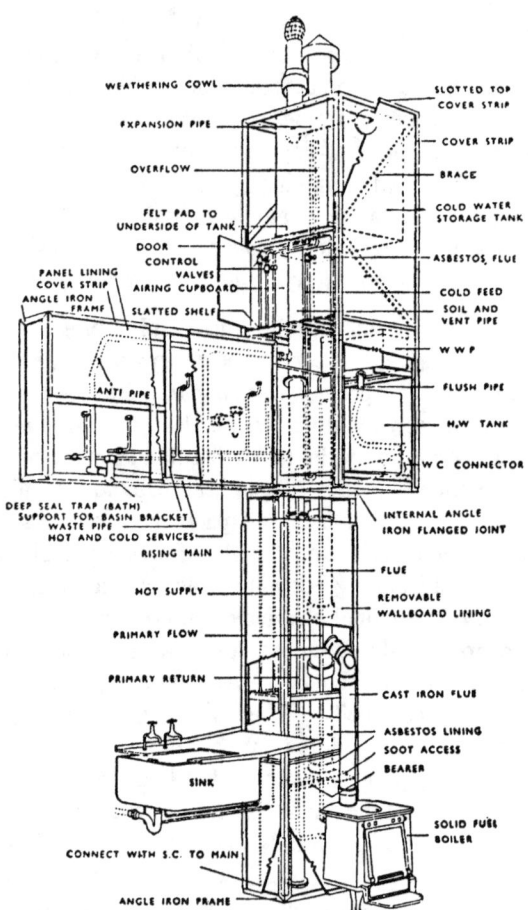

The Denham plumbing and heating system unit, produced by W. N. Froy & Sons, Ltd.
Designers : S. J. Gravely and S. C. Warren.
(*Reproduced by courtesy of W. N. Froy & Sons, Ltd.*)

WILL IT BE EASY TO RUN?

There has been much discussion in the national and technical press about the possibilities of District Heating. It is generally agreed that, given the proper conditions, this method can produce an ideal solution for providing background heating and hot water supply for the individual home. The proper conditions for its use are afforded by closely built-up areas, where many large blocks of flats, terraces of houses, groups of offices and shops are in close proximity. Such a scheme must be worked in conjunction with local authorities so that

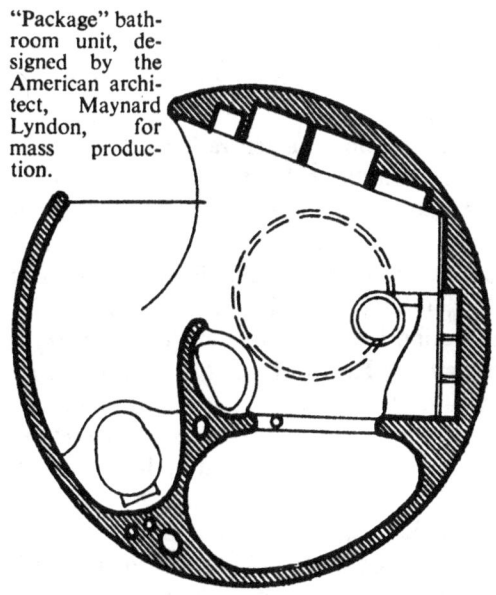

"Package" bathroom unit, designed by the American architect, Maynard Lyndon, for mass production.

proper subways may be constructed under pavements, and, if necessary, leave-ways obtained under buildings themselves to ensure that all pipe-lines are properly protected, insulated and easily accessible.

It has been proved in America, Russia and Germany that on a reasonably large scale, such schemes can be economically self-supporting. Experience in this country has been largely confined to wartime depots, factories and camps when high-pressure hot water has proved a very successful medium for long-distance transmission of heat.

In America oil fuel is more commonly used than solid fuel and steam heat (unsuited to our climate) is often employed. A well known and successful example of district heating in the United States is that of the Parkchester Estate outside New York. This estate covers about 130 acres and contains 12,500 flats in addition to garages, shops, a theatre, community rooms, and so forth, in all fifty-one buildings accommodating 40,000 people. The fuel used is either oil or pulverised coal. Steam is distributed from a central plant to " heating rooms " in twenty-nine separate buildings by means of fully lagged, welded steel pipes laid in concrete box tunnels. In these " heating rooms " water is heated for domestic use by means of calorifiers and the steam for space heating is reduced down to a pressure of 5-lb. per square inch. The cost of heating and hot water per flat is estimated at £7 per annum. Such economy is credited to the use of dry steam and to the good insulation of both the steam pipes and the building structures themselves.

High-pressure hot water (as opposed to steam) is mainly used in Russia, where combined heat and generating stations are common. A condensing steam turbine at a generating station throws away about 55% of the total heat in the coal in the cooling water of the condenser. Actually, power plants only operate at 15% to 25% thermal efficiency. Heat losses occur additionally in the generation of the steam as well as by condensation and leakage. Distribution losses in steam amount to 11% to 15% (about equivalent to electricity) which limits the economic distributing area from one station ; but some of this loss is due to meter faults. Heat is difficult to measure by a meter.

In Great Britain 20,000,000 tons of coal per annum are being burnt in electricity generating stations, of which 11,000,000 tons are wasted for the reasons already mentioned. In addition, 40,000,000 tons of coal are demanded by domestic consumers each year. By organised planning on scientific lines, the power stations could reduce this loss by supplying low-pressure steam or hot water to hundreds of thousands of buildings of various kinds, and this would also save much of the domestic consumption.

These statements were made by Mr. David Brownlie in 1943, and in the article containing them he quoted some historical examples of heating by exhaust steam.[1] The first recorded example is in that of the home (appropriately named Steam Hall) of Matthew Murray, a Leeds engineer who, in the late eighteenth century, heated his house with exhaust steam from his works engine. As a commercial proposition it was first used in the U.S.A. in 1844, at the Eastern Hotel, Boston. It was not a great success owing to insufferable leakage from the pipes when the steam was turned on. In 1877, Birdsill Holly formed a district heating company in Lockport (New York). As a result of its success, Wallace C. Andrews, in 1881, started the New York Steam Corporation which, operating in the Broadway area in

---

[1] *Building*, April, 1943, pp. 100–101.

1882, used three miles of mains. This is now the largest district heating company in the world.

Several British cities have considered incorporating district heating in their post-war housing schemes. A report on a scheme for the Corporation of Bristol for an area of 345 acres shows a very satisfactory estimated return.[1] The estimated total cost, spent over a period of fifteen years, is something over one and a half million pounds, recoverable over a period of twenty-one years from revenue. The annual coal consumption is estimated at 77,500 tons per annum as against an estimated 100,000 tons by normal methods of heating. Considerable electric power is used in such schemes for pumping to produce good circulation. A calorifier system is advocated for water heating in each house, to avoid any extravagant waste on the part of a tenant; which means that each tenant has his own storage cylinder in which the central supply circulates at high temperature through a coil, and when the storage is drawn off a time-lag must occur for heating up the fresh supply. District heating would entail little equipment in the house itself, whether traditionally built or factory-made, but the latter, with its superior insulation, would produce greater economy in heating, and its construction would more easily accommodate radiator and piping services. To be economical, these schemes should serve the majority of building owners in the district heating area. Gas and electricity would still have their fair uses in such areas, both for topping up the background heating, and for cooking.

When the source of hot water supply is within the dwelling, economy of running costs becomes a deciding factor. The assured abundant supply of hot water that comes from district heating passes into the control of the housewife who, before the war, was often struggling to make a small household allowance cover an impossible programme : to her hot water was a luxury : it was an expense that could be and was cut down. Before 1939 the authorities who determined the standards of life for families with low incomes advocated the provision of one solid fuel appliance per dwelling, which was supposed to perform the triple task of space heating for the living room, supplying hot water and facilities for cooking. This maintained the low standard of personal cleanliness which disgraces such a large section of the community, debased still further the already deplorable national standards of cooking, and, by extending the use of the tin-opener, helped to impair the digestions of new generations. It is now recognised that

---

[1] P. G. Kaufman (*Industrial Heating Engineer*, January, 1944) : Summary of conclusions included in report on district heating for Bristol.

no solid fuel appliance yet invented can efficiently perform this threefold function. In most of the housing schemes that have been planned for the post-war period, gas or electric cookers are provided, while a solid fuel appliance serves the living room, with a back boiler for hot water supply. In summer, when the fire is not lit, an electric immersion heater operates in the hot water cylinder or as an alternative an instantaneous gas water-heater is used at the bath or sink. In some schemes, such as those built at Northolt in 1944, an independent boiler has been installed for hot water ; this boiler is supposed to act also as a space heater for the room in which it is placed.

To-day we recognise the significance of efficient insulation, or lagging, for any hot water storage appliance and the pipes that serve it : to omit this protection is to encourage great waste of heat, fuel or power. In low-cost houses built by traditional methods the cheapest place for the chimney breast is against the party wall, and when the fire in this position has to provide hot water as well as heat for a living room, even efficient lagging cannot cut down the heat loss that follows the long journey from the boiler to the storage cylinder and thence to the taps it serves. Although hot water can be provided with reasonable efficiency by the back boiler of a coal fire or from a cooker, the system suffers from the disadvantage of catering for two purposes, when possibly only hot water is needed : it also lacks flexibility, for it determines, perhaps inconveniently, the position of the source for the hot water supply,

Improved coal-burning appliances have been evolved, mainly through the British Coal Utilisation Research Association, and closeable coal fires, which remain alight from some eighteen hours without attention, can provide hot water from a back boiler for early morning use by quick boosting from special controls. Such fires can provide on test four successive hot baths at quarter-hour intervals, fed by a hot water cylinder of some thirty gallons in addition to the boiler itself. It is established that solid fuel is the most economical for hot water, then gas, and lastly, electricity. But in any comparison of costs intermittent use must be taken into account. A steady and low rate of burning as against periods of uneconomical boosting is desirable with solid fuel. Hot water storage systems with gas or electricity thermostatically controlled are dependent for high efficiency on their insulation, and the instantaneous gas heater placed beside a sink, bath or basin proves economical by being entirely suited for intermittent use.

The factory-made house, which is independent of brick and concrete flues, has been evolving in the last few years so-called " kitchen and bathroom units," which embody the heating source of the hot

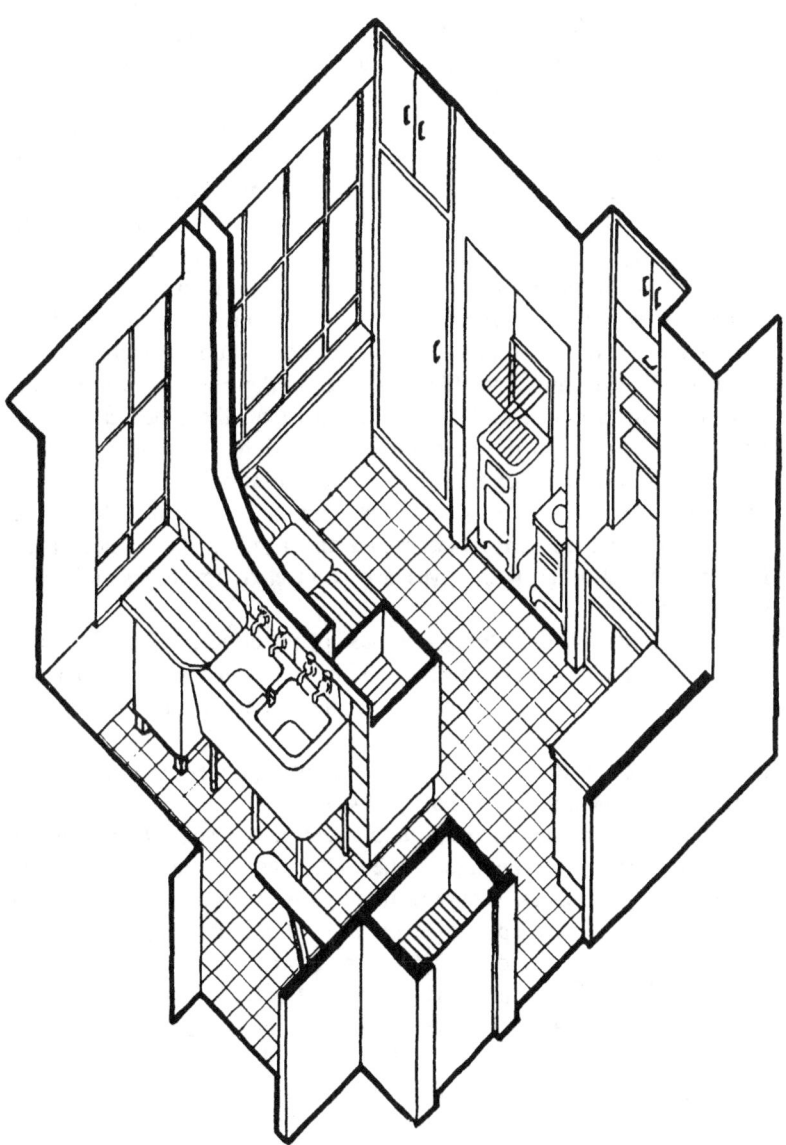

A kitchen working unit, as recommended by the Scottish Housing Advisory Committee on the Design, Planning and Furnishing of New Houses. A utility space for laundry is separated from the kitchen by the plumbing unit.

water, either by an independent appliance or in conjunction with the space heater in the living room, and provide the shortest possible run to all points of supply as well as including the drainage necessary for the various sanitary appliances. These kitchen and bathroom units have been designed both for single-storey dwellings, where the two rooms adjoin each other, and also for two-storey houses, where one room is placed over the other. Such units range from the suggestion made in the Scottish Housing Report for placing the kitchen sink and refrigerator on one side of a plumbing unit, and a double wash tub and a washing machine on the other side, to the far more elaborate and all embracing proposal by the City of Coventry for its post-war factory-made houses. This provides for the living room stove with its back boilers and service to radiators, a sink, cooker and refrigerator with racks, cupboards, and a ventilating hood on the one side, and a wash tub, washing machine, and drying cupboard on the other, all in one steel-framed unit. In the unit immediately above are placed a W.C., bath, basin, linen cupboard and hot and cold supply tanks as well as the flue from the stove. The units contain all the plumbing and service pipes, which can be assembled and fixed to the frames in the factory. The fittings themselves are connected up very simply and easily on the site.

A comprehensive one-storey unit has been produced for the Ministry of Works for emergency housing. This unit, some 10' long, has on one side a cooker, sink, refrigerator and storage cupboards for the kitchen and, on the other, a bath, basin and airing cupboard. The heating stove of the living room with back boiler and hot water cylinder are attached to the end of the unit.

Some units are designed to take independent solid fuel; others gas or electric water heaters. They all have the merit of internal plumbing and of short, economical runs for the supply pipes. Their development has been sponsored by large plumbing firms, by gas and electrical industries, and by joinery and pressed metal firms. This promise of healthy competition suggests that an attractive variety of practical appliances for increasing the comfort of every home will, in due time, become available. Their cost will be determined by their design and the quantity that can be put into production. Their use demands a departure from old type house plans. They are, of course, typical *machine parts* of the factory-made house.

Undertakings concerned with public services such as gas, electricity and water supply provide the same standard quality of product for every type of home : they cater for a universal market, and its size should encourage the mass production of highly efficient well designed appliances. In 1940 there were ten million consumers of electricity in Great Britain. In 1938 the capital represented by the industry amounted to £553,000,000, covering 600 separate authorised undertakings, which varied in scale from those selling a few thousand to five million units a year.[1] With such widespread organisation it is only recently that an attempt has been made towards standardisation in the design and dimensions of appliances, which might reduce costs and increase convenience. There are still too many different voltages and systems of supply, which make private appliances useless when moved from one district to another. There are also far too many different forms of tariff in various parts of the country, some of them even too complicated to be understood by the consumer. Some of the undertakings do not offer adequate hire or hire purchase terms for equipment; others have overlapping rights of supply, with consequent duplication of mains and plant, thus adding additional cost to the consumer.

Despite such conditions, the factory-made house may anticipate a supply of electricity and well-designed and efficient appliances. Excellent work in improving appliances has been done by such bodies as the British Electrical Development Association and by the many progressive manufacturers.

In the past most householders have found little convenience in the layout of electric wiring or of its other essential parts. Fuse boxes, switches, slot meters and meters have frequently been placed in most inaccessible positions and, apart from switches, practically no attempt has been made to improve their design. Even the mechanism of the fuse has proved a nuisance to the householder. In the small house far too few points have been provided. Points for light and heat were kept separate, mainly on account of separate tariff rates with their meters. Far more plug points for appliances and a far higher standard for lighting are not only desirable but essential. The old method of built-in wiring was costly and frequently proved a bottleneck. Surface conduits or flexible wiring in the hollow construction of the factory-made house can improve on this. It is possible that some form of the fluorescent tube may become the universal method of artificial lighting. To install this improvement in millions of traditionally built houses would be a major operation; but the factory-made house can take such improvements at any time with only a relatively small amount of re-equipping.

With a demand for many more plug points to take electrical equipment, advantage would be gained from the " ring " circuit system of

[1] Leslie Hardern, *Architects' Journal*, April 13, 1944, pp. 277–280.

wiring. This could be carried round a room, behind a skirting for instance, and have many points for plug entry, to serve heat or light, without entailing extra wiring. Such a system is made possible by placing a fuse in one of the legs of each plug. The City of Coventry Architectural Department proposes for their housing that this skirting should be of plastic and removable, to facilitate alterations or additions to the outlet sockets.

For simple quick wiring, particularly for factory-made houses with hollow floor spaces, a recent system provides a central ceiling box, which is delivered complete with all its armoured flexible conduit, pre-cut to length and neatly coiled. This box is placed in a convenient ceiling position and the conduit is then joined to the various ceiling points and to switches, which are also placed on the ceiling, and are operated by cords.

Perhaps the greatest revolution in wiring may come from America. The Pierce Foundation has evolved a system of plastic-covered surface wiring which eliminates the customary rubber insulated copper wire and in place uses either copper tubes or copper rods for conduction, with a flat phenolic resin covering. These are made in various lengths for convenience, some with plug outlets and some without, each length being simply socketed into its neighbour. The plastic covering, which can fit neatly into a grooved dado or skirting, does not look unsightly, nor does it collect dirt when used on a wall. It is suggested that, as a system, it is safer than present methods. No leakage is possible, as that which occurs through rubber insulation deteriorating. That it should not prove expensive when marketed is shown by the total cost of the Pierce Foundation prefabricated houses described in Chapter 6, and which were wired by this system.[1]

Electricity is so clean, tidy and convenient; its use for power, lighting and heating seems so appropriate in this commercial machine age, that the far older method of using gas has been overshadowed. But gas for heating and cooking is far from being a back number. It is true that few people would deny the superior convenience and effectiveness of electricity for lighting; but the house, whether factory-made or not, that uses electricity for light and power and gas for heating and cooking, is a well-equipped and comfortable and easily run house. It has been stated by Sir Miles Thomas, Vice-Chairman of the Nuffield organisation, that a high-pressure gas grid is no dream and could be developed as an economic reality, thus providing thermal transport to cure some of the evils due to the intense localisation of

---

[1] See page 95.

industry.[1] The importance of the gas industry is indicated by the estimates of its output. For example, in 1937 it converted £25,000,000 of coal into £75,000,000 of gas, coke, coal tar and other products.[2] In 1937 about 48% of the gas consumed in Great Britain was used on domestic cooking, representing about nine million tons of coal converted to gas and coke.

Great progress has already been made by the lagging of solid fuel cooking appliances, notably with such heat storage cookers as the Aga and the Esse, but the principle of insulation could be applied still more extensively, and to gas and electrical appliances. A test of two electric ovens has been made, one lagged and the other unlagged. The heating-up time of the former was eleven minutes, of the latter thirty minutes. The total saving of the lagged oven over the unlagged was estimated to be between 40% and 50%, or, on a basis of £8 per year running costs, a saving of £3 4s. 0d. per annum.

It is claimed that there is a saving of 20% of gas consumption by insulating the oven.[3] A further possible saving of 20% is claimed by re-designing the hot plate of the gas stove. (On the electric cooker the hot plate is likely to be supplanted by separately heated cooking utensils.) The achievement of such savings depends on the demand on manufacturers to produce improved equipment, and the greatness of the demand determines the lowness of the first cost, if standardised designs and mass production are employed.

The days of the heavy black-leaded cast iron coal range have gone for good. The modern insulated range may be of lightly designed and enamelled cast iron in whole or in part, or of enamelled steel. The range may be designed to give no space heating in the small kitchen, though an insulated fire door may give heat only when required. The gas and electric cooker to-day both tend to be horizontal in design, with the oven at the side of the grill and hot plate. This saves the stooping caused by the low-level oven and allows the apparatus to be built in as a fitment, with cupboards under and on either side.

Cooking appliances for the factory-made house can be easily standardised in dimensions. As factory-made houses will be precision-built, they can accommodate standard dimensions in equipment to a fraction of an inch. There is a general demand that the kitchen of the future should be provided with a considerable amount of equip-

---

[1] In an address to Liverpool business men, quoted in the *Architects' Journal*, November 2, 1944, page 391.
[2] Leslie Hardern, *Architects' Journal*, April 13, 1944, p. 277.
[3] E. W. B. Dunning and R. F. Hayman, " Domestic Gas Ovens," a paper read to London and Southern District Junior Gas Association, February 15, 1938.

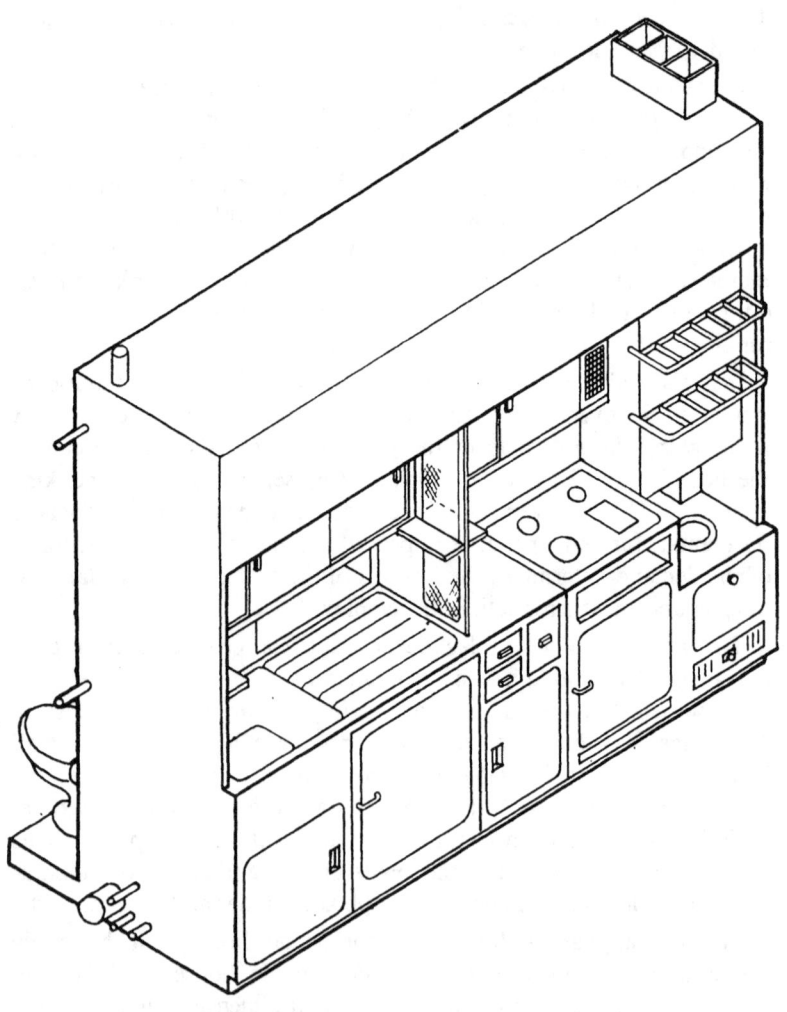

The kitchen side of the Elcock house service unit, embodying sink, draining boards, refrigerator, cooking appliance and water heating unit

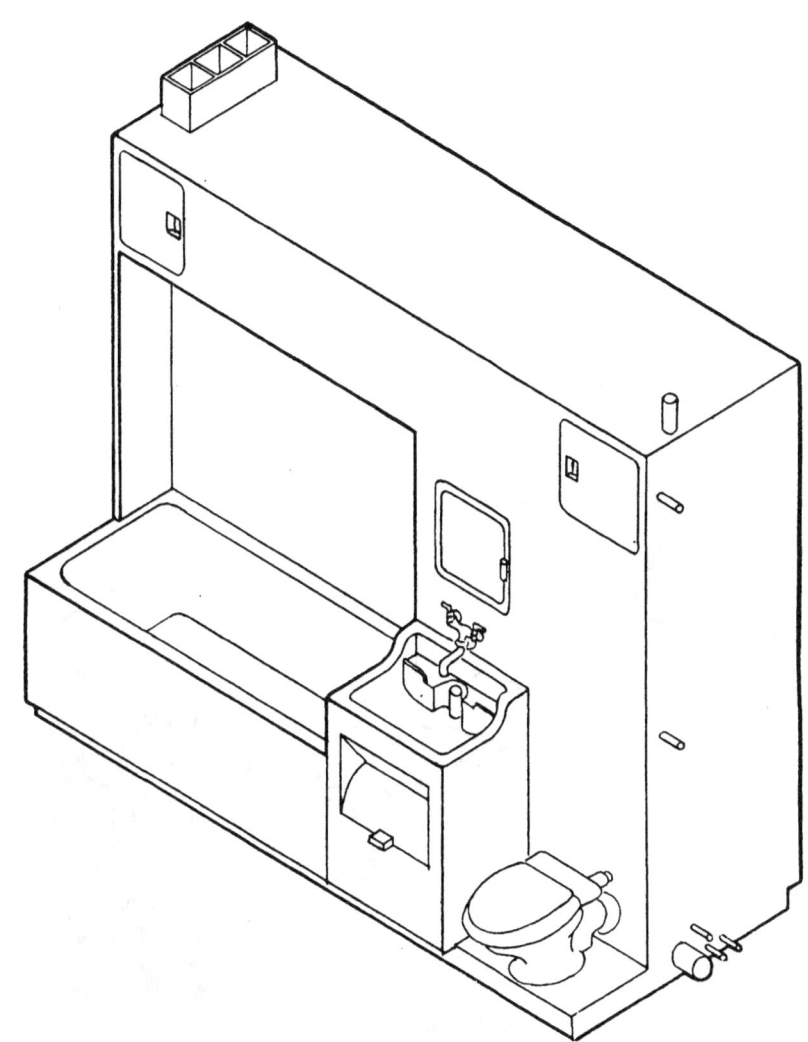

The house service unit designed by the late Charles E. Elcock, F.R.I.B.A. The bathroom side, embodying bath, basin, laundry unit under basin, and w.c.

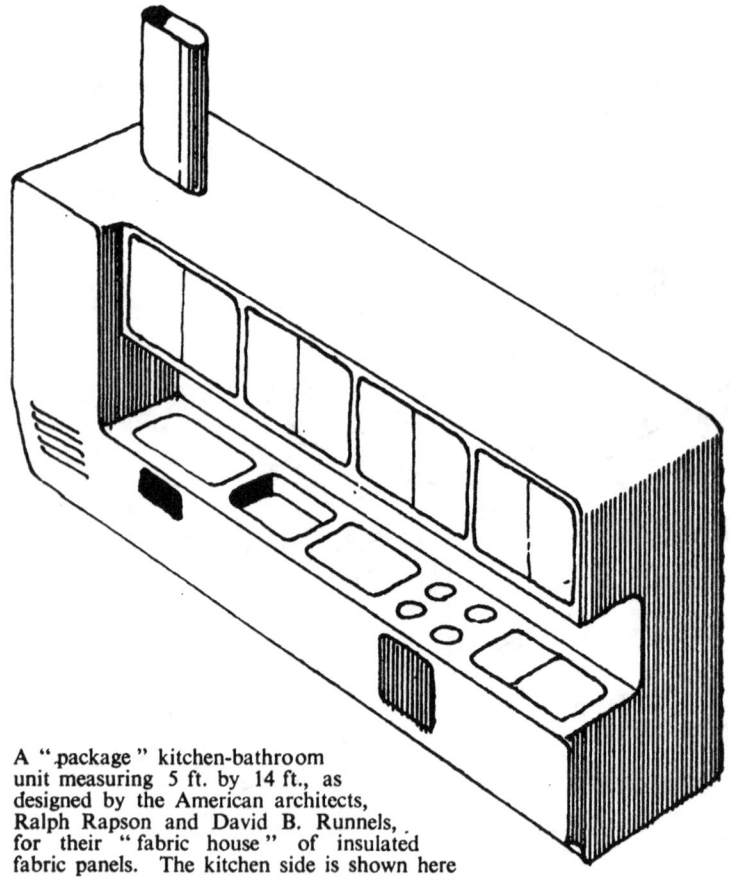

A "package" kitchen-bathroom unit measuring 5 ft. by 14 ft., as designed by the American architects, Ralph Rapson and David B. Runnels, for their "fabric house" of insulated fabric panels. The kitchen side is shown here

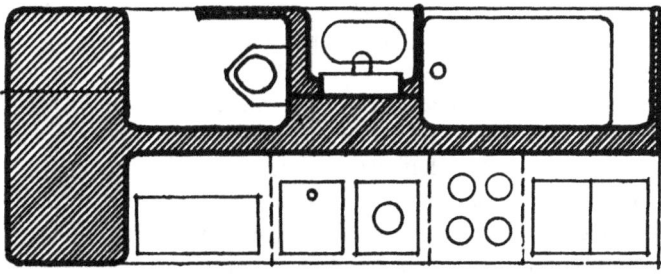

Above: Plan of "package" kitchen-bathroom unit.
Below: Bathroom side

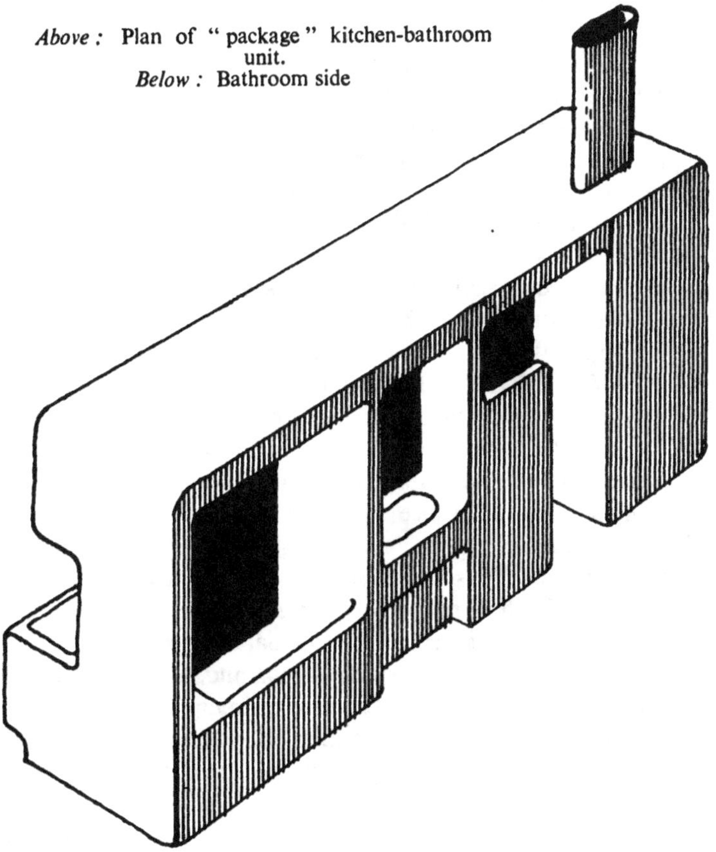

ment for cooking and for storage. Some standard basis for the size of such units is essential to allow proper assemblies to be made up and installed.

Much attention has been drawn in this country to Project A.62 from the U.S.A. This is a proposal to establish a system of modular dimension (four inches has been chosen) on which building and equipment dimensions should be based. The choice of this four inches has been based on the 8"×4" American brick. The project has been prepared by the American Standards Association in collaboration with the American Institute of Architects. The Standards Committee of the Ministry of Works, representing in its membership both the professional and industrial side of building, has given much consideration to this question of establishing a modular dimension. While the Committee has recommended standard dimensions for all kinds of building units and equipment, it does not see a practical application of the American recommendation for Great Britain.

The various systems for British factory-made houses have up to date each fixed the sizes of their standard units according to the demands of the construction and materials adopted. For instance, the wall units widths of the Braithwaite system are 3' 2", of Hill's Patent Glazing 3' 0", of the Uni-Seco 3' 0", of the Key House Uni-built 4' 0". The thickness of internal partitions again varies in a similar way according to their material and construction. There are obviously great advantages to be derived from standardising the dimensions of all equipment that goes into the home, especially into the kitchen, so that allowance for every type of appliance—every machine part—can easily be made. Sink and storage units for kitchens could thus be standardised in size, though marketed in various materials, such as plywood or metal, to give variety of choice, and this choice would also extend to internal arrangements.

Built-in refrigerators will by standardisation and mass production be reduced to far lower cost than was ever contemplated before 1939. An alternative to such food preservation will be a standard cooling unit available for installation in properly insulated standard larders.

The disposal of refuse is still primitive in most of our towns and cities. Within the next half-century we may come to regard the dustbin and the refuse-collecting lorry as we now regard the slop pails and the night carts of the seventeenth and eighteenth centuries—as insanitary anachronisms. Dustbins and refuse containers could be greatly improved in design; but the system is fundamentally wrong, and it should be changed. In the United States refuse disposal, like many other things, has been mechanised. A grinding unit, with a motor, has been designed for attachment to individual sinks.[1] This grinder macerates food waste into pulp, which is then washed down into the drainage system. The waste chamber, of a quart capacity, holds about 1-lb. of food parings and refuse, which when ground are washed down by water at the rate of $2\frac{1}{2}$ gallons per minute. Some 25,000 of these grinders are already in use, and three hundred cities in the U.S.A. have given them a limited approval. The cost of this home grinder is high, and clogging can occur with materials such as pea-pods, and naturally difficulties arise if forks, spoons, and similar objects fall unintentionally into the machine.

It is sometimes asked why refuse cannot be burnt to provide fuel for district heating. The answer is, that it takes the refuse of about seventy houses to provide heating and hot water for one house.[2] The refuse varies, and is extremely uncertain in its calorific properties; at best it could only act as a subsidiary to other fuel.

A well established method of refuse disposal applicable to large blocks of flats is the Garchey system; but, since the refuse is washed through the sink into a central collecting chamber, it would not prove economical for dwellings other than flats. The individual house— even the factory-made house—may have to await new technical advances before a refuse disposal system comparable in efficiency with other services can become generally available.

From this highly condensed survey of services that are now available for the factory-made house, the question: " Will it be easy to run ? " may be answered with an emphatic *yes*! It would be truly labour-saving, because good design and the application of technical knowledge have together expelled the drudgery that has haunted the housewife for centuries.

---

[1] *Post-war Building Studies*, No. 9, Mechanical Installations, paragraphs 156 and 157. H.M. Stationery Office, London, 1944.
[2] " Utilisation of Waste Heat from Refuse Disposal Works," by E. S. Hobson. *Heating and Ventilating Engineer*, February, 1944, p. 313.

CHAPTER 9

What Will it Look Like?

WHEN the small houses of the Georgian period that adorn the country were first built they caused much unfavourable comment—even alarm—because their raw red brick was considered hideous. Landlords great and small were prepared to plant trees and lay out parks for the benefit of posterity, but only the wealthy owners of estates could afford to build in stone, which was deemed the proper material for the mansions of the nobility and gentry. (" Brick, Mr. Pitt, brick ! " snapped George III when he was shown the country house of the great statesman ; and Mr. Pitt was so abashed that he had the walls faced with stone before another royal visit took place.) People were prepared to wait for trees to grow, even though they knew they would not live to enjoy the fullness of their shade ; but they were not so patient about building materials that needed time to acquire the mellowness that we now admire. Still, they knew that bricks and tiles, stone and timber, would acquire beauty with the passage of years, and that such beauty would be cumulative, for the best traditional materials " weather " in the most agreeable fashion. To-day we are not so certain that our contemporary domestic architecture will age graciously, and may wonder if, apart from composition and architectural features, traditional building offers much variety in external finish.

Brick walls in the majority of housing schemes are still harsh and unsympathetic to the eye, years after erection. Cement rough cast has been a popular alternative, and a coloured plaster finish has been used sparingly in order to avoid upkeep. Neither tile, slate nor asbestos has been able to provide roofs of a quality that could achieve beauty by weathering ; to-day roofs of such materials look as hard and new in many a housing scheme as when first put up.

Traditional materials in the hands of the speculative builder all too often achieve the monotony that is dreaded by the critics of the factory-made house. Again we may ask, how well have the four million houses built between the two world wars with traditional methods and materials done their job ? With some sixty years of life still before them, very many are already out of date in planning and equipment. Moreover, only a small percentage of them represent architecture at all. They are untouched by the beauty of our traditional domestic architecture and its materials. Pitched roofs of machine-made clay or composition tiles and walls of wire-cut bricks can never, even with the flight of centuries, produce the mellow beauty of houses in a Cotswold market town, like Chipping Campden, or an ancient port like King's Lynn. The festoons of outside plumbing, so impractical in frosty weather, are a constant eyesore, and the pretentious imitations of past architectural styles carry neither conviction nor charm.

The factory-made house cannot, if its materials are used effectively, imitate any traditional architectural style. As a structure it does not, in the sense of individual design, come into the category of " architecture " at all ; though the disposition, the siting and layout of factory-made houses are essentially architectural problems. Factory-made houses, even with a life expectation of fifty years, are in the nature of consumable goods. They represent a departure from all hitherto accepted ideas about building. They have nothing to gain from time. Age and exposure to weather will not produce the attractive wrinkles and apple cheeks of healthy old age. The choice of their materials is influenced by the knowledge that nature is an enemy rather than a friend—an enemy searching for weak spots, who must be defied and defeated. Nevertheless, many architectural qualities may be maintained by the new building technique, namely, good composition and proportion, good colour, and good design in every detail. Externally, the factory-made house can be clad in a variety of materials. If a metal cladding is used it can be cheerfully painted ; and those who have fears for the landscape should ask themselves whether they have ever been offended by white-walled or colour-washed cottages or houses, when they are pleasantly disposed on a site. The disposition of the houses on the site ; the judicious variation of the units when they are assembled, and an imaginative use of colour can dispel the bogy of monotony. Again, the use of composite materials defeats monotony, and we may repeat that point made in Chapter 2, that designers no longer think in terms of " All Steel," " All Concrete," or " All Timber " houses. Variety in external finish does not diminish the advantages of factory production ; and we may also repeat that " several systems allow an outer brick skin to be built, to accommodate the taste and secure the confidence of those who still cling to ' bricks and mortar.' "[1] But it is more likely that non-traditional materials would be used for large quantities of factory-

[1]Chapter 2, p. 16.

made houses, and as they serve the needs of mass housing, their design happily would seldom be applicable to single isolated dwellings. Such individual houses must " grow " into their surroundings ; and in the past their affinity with mother earth has been achieved by the use of local materials and building methods. For example, in the Cotswolds, " The houses seem part of the earth ; they crouch comfortably rather than stand, and the impression of low length is strengthened by the placing of the windows—they nestle up into a gable as part of it ; dormers sit low on the eaves as if seeking the support of the wall below. The chimneys rise out of the ridge with a comfortable suggestion of supporting the roof."[1]

Modern country house architecture approaches this harmonious partnership with its environment by the use of walls and terraces and other devices of the garden designer, which may be said to stitch the composition into the tapestry of the landscape ; but this demands costly site work, and money for this would not normally be available for one or two cottages. The opportunity for good architectural composition will be furnished by factory-made dwellings when they are used in large numbers, with their appropriate road layouts, gardens and connecting walls and fences. The recommendations for mixed communities, and for communal buildings to serve them, would, if put into effective practice, relieve housing estates of the monotonous features so often associated with large scale schemes of traditional domestic building.

The modern conception of road layout makes for flexibility, not rigidity. The speculative builder nearly always thinks of houses in rows on identical plots, and thus makes an open-work version of the old, evil, back-to-back housing of the early nineteenth century industrial slums, a back-to-back-garden pattern that is wholly destitute of imagination, convenience, charm or—most important of all in Britain—privacy. Factory-made houses need not be regimented in rows ; they can be given both beauty and variety of setting.

For example, the small cul-de-sac, the double driveway, with its grass divisions and its well placed trees, the intimate footpath, and the small service road, all have their place ; all can be planned in the interests of convenience and variety. Natural features, existing trees, can be used to the best advantage. The general reaction of the speculative builder who acquires a piece of land that is " ripe for development " is to cut down every tree in sight, and to use such mechanical devices as the bulldozer to level out the site ; but the

---

[1] *Men and Buildings*, by John Gloag (Country Life, Ltd., 1931). Chapter IX, " The Structural Revolution and the New Materials," p. 165.

imaginative planner would use the bulldozer either for making hills or for sweetening out levels. He would practise what in America is known as "landscaping," and this artificial revision of sites has given to thousands of localities in the United States agreeable, secluded and beautiful settings for housing schemes.

Unlike the property owners of the eighteenth century, we need no longer await the last few years of our lives to see the results of tree-planting: forestry research has greatly increased the possibility of transplanting even quite large trees from one site to another. One of the most remarkable demonstrations of tree transplanting was afforded by the San Francisco World Fair in 1939, which was built on an artificial island made in San Francisco Bay. Although " Treasure Island," as it was called, was used as a site for the Exhibition, it was made in order to establish a permanent air base for the China clippers. To enhance the attraction of the island as an exhibition ground, large trees were given by private owners in different parts of California, transplanted for the duration of the Fair, and then returned to the owners. The cost of transplanting trees is high when they are large, sometimes reaching a figure of £200 a tree. Except for special features in a housing scheme, this sort of transplanting would be too expensive. But it should be remembered that trees are infinitely preferable to the so-called architectural adornments that disfigure so many localities: the Victorian monuments, and—worse—the statuary of that period.

The blocks and terraces which factory-made houses may produce, will at first seem unfamiliar in character, and it will be the natural features of the landscape, trees, lawns, hedges and bushes, which will convey the friendly character that will soon make the setting as familiar, acceptable and far more pleasant than the rows and rows of semi-Tudor, semi-Swiss creations of the speculative builder. What great designers of landscapes, like " Capability " Brown, did for the great estates of the nobility and gentry in the eighteenth century our town planners and architects could do to-day for our towns and villages, for the benefit and enjoyment not only of the residents, but of the whole people. The factory-made house, compact and tidy in design, finished in colours that brighten life instead of depressing it, like the dark terra-cotta paint—" dried blood " as it has been called— and the mournful green that are so often used because they are supposed to be " practical," could restore to the town and countryside something which has been missing since the Middle Ages in domestic architecture—gaiety.

CHAPTER 10

Consider Your Verdict

EACH chapter of this book has asked a question and has attempted an answer. Here the conclusions of those nine chapters may be briefly summarised.

We need the factory-made house because we must get rid of our slums, replace our damaged houses, and provide new and better homes for future generations. We must be quick, we must be cheap, we must be efficient in solving the housing problem. The factory-made house would bring to its solution our industrial strength, knowledge, experience and inventiveness.

An abundance of suitable materials exists which could be economically and effectively used for the industrial production of houses. The economic laws of mass production would allow savings to be made in the handling of materials, in the amount of materials actually used, and in the man-hours needed for erecting houses. Such savings would represent a secret subsidy, which would allow far better standards of accommodation and equipment to be available for everybody—better than most people have hitherto enjoyed or even imagined. In order to take advantage of the economic laws of mass production the quantity of factory-made houses should be guaranteed by the State, acting through local authorities. It might be laid down that 25% of any local housing scheme could be built by traditional methods—which would satisfy regional building interests—while 75% must consist of factory-made houses, and that the sites for the latter must be prepared concurrently with sites for traditional types. This would allow for the erection of factory-made houses immediately after delivery. Given such an assured market, a new house manufacturing industry could greatly reduce the individual cost of dwellings and make them at high speed. Parts for factory-made houses could pour off the production line, ready for assembling on properly planned and prepared sites throughout the country, at the rate of many thousands a week, until we had made up the huge housing deficit that confronts the building industry, the country and the people. The satisfying of this known and measurable market would enable the houses to be produced so cheaply that State subsidies might be quite unnecessary.

This book has been concerned only with factory-made houses that would have a reasonably long life. The purely temporary ten-year type of factory-made house has been regarded as uneconomic and undesirable. The factory-made house could and should last for half a century, it could and should be extremely comfortable, for its cost will be low enough to enable all kinds of appliances to be used which will make for smooth running and will reduce housework.

It would be designed to be labour-saving in the basic sense, and would make the best use of modern services, so that heating, lighting and cooking would be simplified, cheapened and improved. In appearance it would be far pleasanter, and far less likely to create an effect of monotony than the bungalows and jerry-built terraces and semi-detached houses which now deface our residential areas and are the result of the speculative builders' ignorance, incompetence and greed, which cause him to mishandle traditional building materials and methods.

The factory-made house is no vague promise of a paradise to come; it is technically possible, here and now. Industry can satisfy the motorist; industry could satisfy the householder; we could have good housing in our time; finish off the job of slum clearance that we had tackled slowly but thoroughly before the second world war; and create a new house manufacturing industry which would not only serve a great market at home, but might help to solve the housing problems of all the devastated countries of Europe.

The facts have been given here in outline, but a great body of published information about factory-made houses and the principles of prefabrication exists, both in this country and the United States. Every statement we have made about systems of prefabrication, materials, equipment and services may be checked by those who desire to extend their study of the subject. Detailed information has appeared in the following technical journals :—

The Architect and Building News.
The Architects' Journal.
Architectural Design and Construction.
The Architectural Review.
The Builder.
Building.
The Journal of the Royal Insitute of British Architects.
The Official Architect.

Published in Great Britain.

The Architectural Forum.
The Architectural Record.
Pencil Points.

Published in the U.S.A.

A Report, entitled "A Survey of Prefabrication by the Ministry of Works," by D. Dex Harrison, A.R.I.B.A., A.M.P.T.I., J. M. Albery, A.R.I.B.A., A.M.P.T.I., and M. W. Whiting, A.R.I.B.A., which was the result of three years' detailed research on methods of prefabrication, was issued in March, 1945, and was distributed to various libraries throughout the country in the summer of 1945.

This book will be only one of many on the subject; but it has been written in order to put the facts before those who will ultimately decide the case for or against the factory-made house—the British public and their elected representatives, in Parliament and Local Government. So weigh the evidence, and consider your verdict on the factory-made house.

The need is urgent; the factories are there; the materials are available; the technical skill exists; the task will provide thousands of jobs for thousands of men and women who have kept our country free and fit to live in.

Who will prevent the people of Britain from having the homes they deserve?

THE END

INDEX

Index

Adhesives, 5
Aggregates, for concrete, 11
Aircraft Industries Research, 4, 66
Airey, Sir Wm., 44
Airform, 17
Akron (Ohio), 65
Aladdin Bay, 55
Alloys, light, 14, 66–7
All-Steel house, 9, 15
Aluminaire house, 76–8
Aluminium, 14
  ,, alloys, 66–7
  ,, Union Limited, 67, 69–71
America, steel houses, 9, 23–5 ; timber houses, 11, 55–6, 59–65 ; concrete houses, 44, 52 ; consumer needs, 84–5 ; district heating, 117–18
American Dwellings Corporation house, Pl. 7
Andrews, Wallace C., 118
Appearance, of factory-made house, 133 ff.
Architects, and prefabrication, 81
Architecture, 133
Arcon house, 35, 41–3, Pl. 23–5
Armco panels, 23, 27
Atholl, Duke of, 9
  ,, house, 20, Pl. 4–5
Atmosphere, pollution of, 107
Attebury, Grosvenor, 44

Back-venting, 113
Bates Prefabricated Structures, 65
Bathroom unit, package, 117
Baths, introduction, 111
Battersea, 98
Bauhaus, 75
Beaucrete system, 42–4
Bernal, J. D., 104
Birmingham, 35, Pl. 17–22
Bogardonson, James, 15
Boots system, 42, 44
Boswell, James, 4
Boughton, H. V., 4
Braithwaite and Co., 24, 29, Pl. 4

Braithwaite house, 25, 31–4, Pl. 36–9
Bramah, Joseph, 111
Brick skins, 16
Bristol, 119
British Cast Iron Research Association, 103
British Coal Utilisation Research Association, 103, 107, 120
  ,, Electrical Development Assocn., 123
  ,, Iron and Steel Federation, 35, 47, Pl. 29–31
  ,, Steelwork Association, 6
Brown, " Capability," 135
Brownlie, David, 118
Bubble houses, 19
Building Research Station, 103
Burt Committee, 105

Carolina, Pl. 3
Cast iron, 14–15
  ,, houses, 67
Celotex, 12
  ,, Corporation, Pl. 8
Churchill, Winston, 103
Cladding, 133
Clapboards, 3
Clothed Concrete Constructions, Ltd., Pl. 14
Coal consumption, British, 118
  ,, fires, 107
Colnbrook, 44, 49–50, Pl. 15
Composite houses, 71–5
  ,, production, 9, 16
Composition, architectural, 134
Concrete, 10–11
  ,, systems, 42–53
  ,, wall, positioning, Pl. 8
Consumer needs, attention to, 83
Convection of heat, 108
" Converted " houses, 87
Cook, Sir Wm., 106
Cookers, heat storage, 125
Cooking appliances, 125–30
Corbusier, le, 80

Costs, 7 ; of council house, 93 ; traditional building, 94 ; proportion to floor area, 94 ; and time-saving, 94 ; sample, for prefabricated house, 97 ; labour, rise in, 100 ; materials, rise in, 99–100
Cotswolds, 133, 134
Coventry, 4, 26, 27, 35, 111, 122, Pl. 33
Cowieson house, 20
Crystal Palace, 14
Cummings, Alexander, 111 63-5

Datchet, 48, Pl. 26
da Vinci, Leonardo, 109
Demounting, 61
Denham plumbing and heating unit, 116, Pl. 48
Dent and Hellyer, Ltd., 114
Design, 88-9 ; and costs, 95
Dimensions, standard, 130
Dorlonco house, 20
Douglas, Wilmot C., Pl. 8
Douglas Fir plywood, 12
  ,,   ,, Plywood Association, 55, 63-5
Douglas house, Pl. 5
Drancy, 52
Drew, Jane, Pl. 46-7
Dry building methods, 2, 6
Dundee, 55
Duo-slab system, 42
Dyke house, Pl. 14
Dymaxion house, 71, 109–11

Edinburgh, 4, 27
Edward I, 107
Egerton, Alfred, 105-6
Elcock service unit, 126-7
Electrical equipment, 123 ff.
  ,, supply, 123
Elizabeth, Queen, 86
Equipment, fitting of, 2
Evans, Oliver, 109
Evelyn, John, 107
Exhaust steam heating, 118
Extension, of factory-made houses, 90

Factory-made house, advantages, 5-6 ; prejudice against, 80
Federal Housing Administration, 71
Ferrie, J. H., 53
Fidler system, 42
Finish, external, 132 ff
Fires, open, 106
Fitzmaurice, R., 107

Flues, 108
Foam slag, 11, 53
Ford, Henry, 79, 91-2, 100–1
Fordite, 92
Forest Products Laboratory, U.S.A., 56
  ,,   ,, Research Laboratory, 103
Forssjo, 65
Fort Wayne, Pl. 6
Frame house, 3, 55
France, concrete houses, 52
Frey, Albert, 78
Friberger, Eric, 66
Froy, W. N., and Sons, 116
Fuller, R. Buckminster, 22, 23, 71, 109
Fulmer, O. K., 78

Garchey refuse disposal system, 131
Gardeners, and heating, 106
Gas industry, 125
Gas supply, 124
General Housing Corporation, Pl. 12
George III, 132
Germany, concrete houses in, 44, 51 ; steel houses in, 9, 25
Gibberd, Frederick., 35, 42, 47, 48, Pl. 26-7, 29
Gibson, Donald, 112
Glasgow, 53
Gravely, S. J., 116
Grinders, refuse, 131
Gropius, Walter, 75
Grundy (Teddington) Ltd., Pl. 42
Gunnison, Foster, 84
  ,, Housing Co., 55, 63, 84

Harington, Sir John, 111
Hawthorne, Nathaniel, 87
Heating, 104
Heating methods, efficiency, 106
Heat leakage, 104
Hendon, 32, Pl. 36
Hills Patent Glazing Co., 32-5, 36-40, Pl. 16, 17
Hodgson, 55
Hodgson (E. F.), Co., 56
Holly, Birdsill, 118
Homasote, 5, 12, 61-3, 65
Hot water supply, 115-20
Howard house, 42, 48, Pl. 26-8
Howard, John, and Co., 42
Hull, Jonathan, 109

Ibo system, 55
Insulation, 6, 83 ; thermal, 104

INDEX

Iron, waste by rust, 14

Jicwood house, 56–60, Pl. 40–1
Johnson, Dr., 4

Katona, E., 49–50, Pl. 15
Keyhouse Unibuilt house, 26, 27–9, 30
Kitchen and bathroom units, 121 ff.
Kitchen units, 126–30
Kocher, A. L., 78
Kvarnholmen, Pl. 2

Labour, rise in costs, 100
Labour-saving equipment, 102 ff.
Lagging, of cookers, 125
Landscaping, 135
Lee, Donovan H., 35, 47, Pl. 29
Leeds, 44, 52
Littlewoods, Ltd., Pl. 46–7
Lochend, Pl. 5
Loewy, Raymond, 75
London County of housing statistics, 97–8
Los Angeles, 56
Lumber, metal, 25
Lyndon, Maynard, 117

Machine-core house, 72
Machine production, principles, 80
Magnesium, 14
Manzoni, H. J., Pl. 17
Mast-hung houses, 71, 72, 109 ff.
Materials, choice of, 15–16 ; rise in prices, 99–100 ; traditional, and appearance, 132–3 ; use in association, 13
Metal sections, pre-cut, 25
Middle classes, housing for, 82
Minoans, 111
Monocoque house, 13
Monotony, 5, 133
Mopin, 25, 52
Mottram, A. H., 19
Moulds, for concrete construction, 10
Munich, 51
Murray, Matthew, 118

National Homes, 65
National Physical Laboratory, 103
Naugle house, 23
Neff, Wallace, 19
Neutra, R. J., 78
Newton Chambers and Co., 67
New York Steam Corporation, 118
Nils Poulson house, 23
Northolt, 32, 36–40, 100, 120, Pl. 16
Northport, 76–8

Oakland (Calif.), 65
Old houses, disadvantages of, 86
One-pipe plumbing, 113
Orlit system, 42, 49–50, Pl. 15
Overcrowding, 98

" Package " bathroom unit, 117
   ,,   kitchen-bathroom units, 128–9, Pl. 46–7
Packaged system, 75
Parkchester estate, 117
Parts of house, number, 4
Paxton, Joseph, 14, 107
" Performance " of house, 83–4
Permanence, 81
Pierce, John B., Foundation house, 95, 96, 124
Pitt, William, 132
Plastics, 12–13 ; uses of, 13
Plat, Hugh, 106
Plumbing, costs, 93
   ,,   units, 111 ff., Pl. 25, 28, 39, 42, 48
Plywood, 11, 56 ; houses, Pl. 6, 7 ; resin-bonded, 13
Portal, Lord, 99
Portal house, 23, 89, 93, 108
Post-stressing, 53
Poulton, Dennis, 6
Precision methods, 2
Prefabrication, 2
Presweld, 32
Prices, rise of, 99–100
Privacy, 81
Project A-62, 130

Rapson, Ralph, 129
Refrigerators, 130
Refuse disposal, 130–1
Resins, plastic, 12, 13
Robertson units, 27
Rockefeller Center, 66
Roof, flat, 105
Runnels, D. B., 129
Russell, R. D., 67, 69
Russia, 99 ; district heating in, 118

San Francisco, 135
Scandinavia, 2–3
Scanhouse, Ltd., 54
Scano house, 54, 55
Scherrer house, 25
Scientific and Industrial Research, Department of, 103, 107
Scotland, 9 ; timber houses in, 11

Scott, Sir Giles Gilbert, 14
Scottish Housing Report, 121–2
Sears Roebuck (Pre-cut) house, 56
Second Scottish National Housing Co., 19
Seco-Unit house, Pl. 13
Shared dwellings, London, 97–8
Shaw, G. B., 115
Sheppard, Richard, 26, 56, 57, 60, Pl. 32, 40
Shoreditch, 98
Short-lived houses, 89
Sjostrom, C., 97
Slums, 1
Smith, Percy, 88
Smoke emission, 107–8
Sound insulation, 83
Speed of erection, 6
Springboig, Pl. 5
Spring-Rice, Margery, 98
Steel, 8–10
Steelox panels, 23
Steel systems, 19–42
Steel Utilisation, Technical Bureau for, 27
Stran steel, 25
Stressed skin construction, 56
Sulston, V., 99
Sweden, concrete houses, 53 ; timber houses, 55, 65–6, Pl. 2–3
Systems, partial and complete, 17 ; shell and skeleton, 17

Tait, Thomas, Pl. 14
Tampa, 52, Pl. 8
Tarran Industries, 55
Telephone box, 14
Telford house, 24, Pl. 4
Temperatures, room, 105
Thomas, Sir Miles, 124
Thorncliffe system, 67
Timber, use in Scandinavia, 3 ; combination with steel, 10 ; modern use, 11–12 ; grading, 12

Timber Development Association, Pl. 43–5
Timber Hill, Pl. 13
Timber houses, Pll. 1–3
Timber systems, 55–66
Tingay, John P., Pl. 43–5
Tipton Green lockhouse, 67
Trailer house, 69–71
Trees, use of, 135
T.V.A., 59, 61–4 ; houses, Pl. 9–11

Unibuilt houses, 4
Unions, Trade, and labour disputes, 9
Uni-Seco type, 72–4, Pl. 13

Van Ness house, 65
Venning, H. J., 94, 100

Wallboards, 5
Wall unit, size, 5
Warren, S. C., 116
Water absorption of walls, 104
Water-closet, 111
Water heating, 119
Watling Estate, Pl. 4
Weatherboards, 3
Weight of house, 4
Weir, Lord, 9, 23
Weir house, 9, 20, 21
Welding, aluminium, 14
Wells, H. G., 87–8
Weybridge, 56, 57, 60
Winget system, 42
Wiring, electrical, 123–4
Working-class housing, 82
World Fair, American, 15, 135
Wornum, Grey, 26, Pl. 32
Wren, Sir Christopher, 80
Wyatt, James, 107

Yardley, Pl. 4
Yorke, F. R. S., 31, 33, Pl. 36
Young, Owen D., 84

PLATES

PLATE 1

Above: Old timber houses in Carolina, U.S.A.
(*Photograph, F. R. Yerbury.*)

Below: Typical timber house in Sweden, of the inter-war period.
(*Photograph, F. R. Yerbury.*)

PLATE 2

Above : Terrace of family timber houses for workers at Kvarnholmen, the Co-operative Society's industrial centre outside Stockholm.
(Photograph, F. R. Yerbury.)

Opposite : Swedish prefabricated timber houses in course of erection, and the completed house.
(Photograph, F. R. Yerbury.)

PLATE 3

PLATE 4

Above : General view of Atholl houses built at Watling Estate, Edgware, London.
(*Reproduced by courtesy of Atholl Steel Houses, Ltd.*)

Below : Telford steel houses, produced by Braithwaite & Co., Engineers, Ltd., erected at Yardley, Birmingham, in 1925-6.
(*Reproduced by courtesy of Braithwaite & Co., Engineers, Ltd.*)

PLATE 5

Above : A pair of Weir houses of the " Douglas " type, built at Springboig, Glasgow, in the inter-war period.
(*Reproduced by courtesy of the* Architect & Building News.)

Below : A pair of inter-war period houses erected on the Atholl principle, at Lochend, Edinburgh.
(*Reproduced by courtesy of Atholl Steel Houses, Ltd.*)

PLATE 6

Plywood house produced for the Fort Wayne Housing Authority at an estimated cost of $1,300.

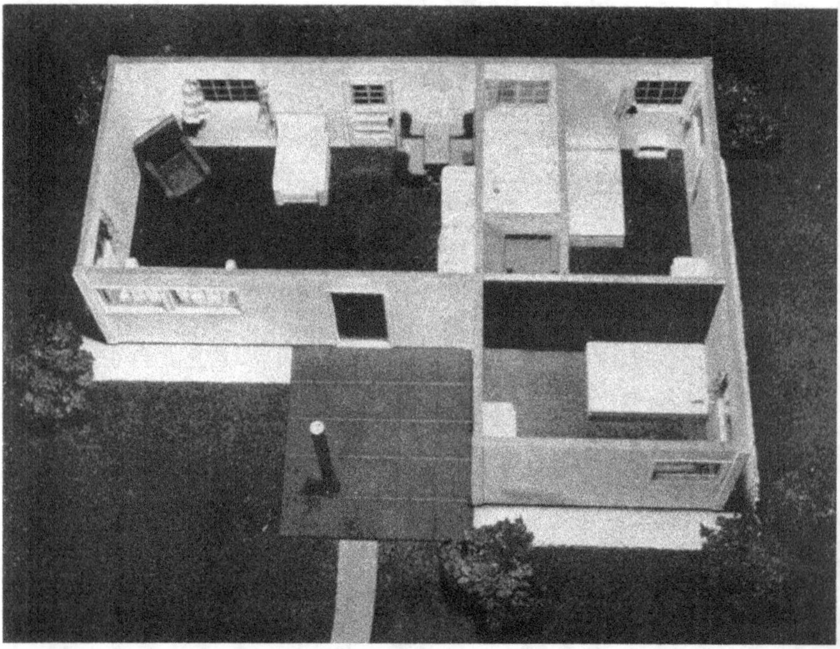

PLATE 7

Houses produced for the American Dwellings Corporation, in phenol resin-glued plywood. *Above:* elevation.

To the left: Interior of living room.

(Reproduced by courtesy of the Architectural Forum *of New York.)*

PLATE 8

Above: Wall of 2½-in. thick concrete, reinforced with pre-stressed steel, being placed in position four days after pouring. Houses built in this manner at Tampa, Florida, U.S.A., were erected in the course of a large war housing project, at a cost no higher than those used for prefabricated wooden houses. With this method of casting concrete walls there is practically no forming.
(*Reproduced by courtesy of Vacuum Concrete Inc., Philadelphia, Penn., U.S.A.*, who hold patents in Great Britain for the vacuum process and electric pre-stressing.) (See page 52.)

Below: House designed for the Celotex Corporation, Plainfield, N.J., U.S.A., by Wilmot C. Douglas, Architect ; Consultants : Harrison & Fouilhoux.
(*Reproduced by courtesy of the* Architectural Forum *of New York.*)

PLATE 9

The T.V.A. sectional house in transit and in course of erection. The completed house and other examples of the T.V.A. sectional houses are shown on plates 10 and 11.

PLATE 10

Above : The T.V.A. sectional houses assembled. See plate 9 for illustrations of house in transit and in course of assembly.

Below : A group of T.V.A. sectional houses.

PLATE 11

Above: Street of T.V.A. sectional houses.

Below: Close-up of the porch.

(*The illustrations on this plate and on plates* 9 *and* 10 *are reproduced by courtesy of the* Architectural Forum *of New York.*)

PLATE 12

General Housing Corporation's multi-units. *Above:* **The** completed half house being moved out for transport to the site. *Below:* Two halves in process of being bolted together to make a house 24 feet wide by 32 feet long.

(*Reproduced by courtesy of the* Architectural Forum *of New York.*)

PLATE 13

Experimental pair of demountable cottages on the Seco-Unit System, erected at Timber Hill, Surrey, by Uni-Seco Structures, Limited. (See pages 72, 73 and 75.)

(*Reproduced by courtesy of Uni-Seco Structures, Ltd.*)

PLATE 14

Above: A prototype house in course of erection at Stoke-on-Trent, produced by Clothed Concrete Constructions, Limited, on the Dyke system of construction. Architect: Thomas Tait, F.R.I.B.A. Consulting and Designing Engineers: Considere Constructions, Ltd.

Below: The completed prototype house at Stoke-on-Trent.
(*Reproduced by courtesy of Clothed Concrete Constructions, Limited.*)

PLATE 15
Pair of Orlit concrete houses built at Colnbrook, Middlesex.   Architect : E. Katona.
(See pages 44, 49 and 50.)
(Reproduced by courtesy of Orlit, Limited.)

PLATE 16
Concrete clad houses at Northolt, produced by Hills Patent Glazing Company, Ltd.
(See pages 32, 35, 36 to 40).
(*Reproduced by courtesy of Hills Patent Glazing Company, Ltd.*)

PLATE 17

City of Birmingham experimental houses constructed by Hills Patent Glazing Company, Ltd. Architect: Herbert J. Manzoni, City Engineer and Surveyor. The kitchen interior is shown below.

PLATE 18

Above : City of Birmingham experimental houses in course of erection. (See plate 17.)

Below : View of steel framework showing steel units fixed in position by clips fitting into lugs welded to the stanchions.

Opposite, and on plates 20, 21 and 22 are shown details of City of Birmingham experimental houses.

(The illustrations shown on plates 17 to 22 are reproduced by courtesy of the City of Birmingham Public Works Department.)

(See page 35.)

PLATE 19

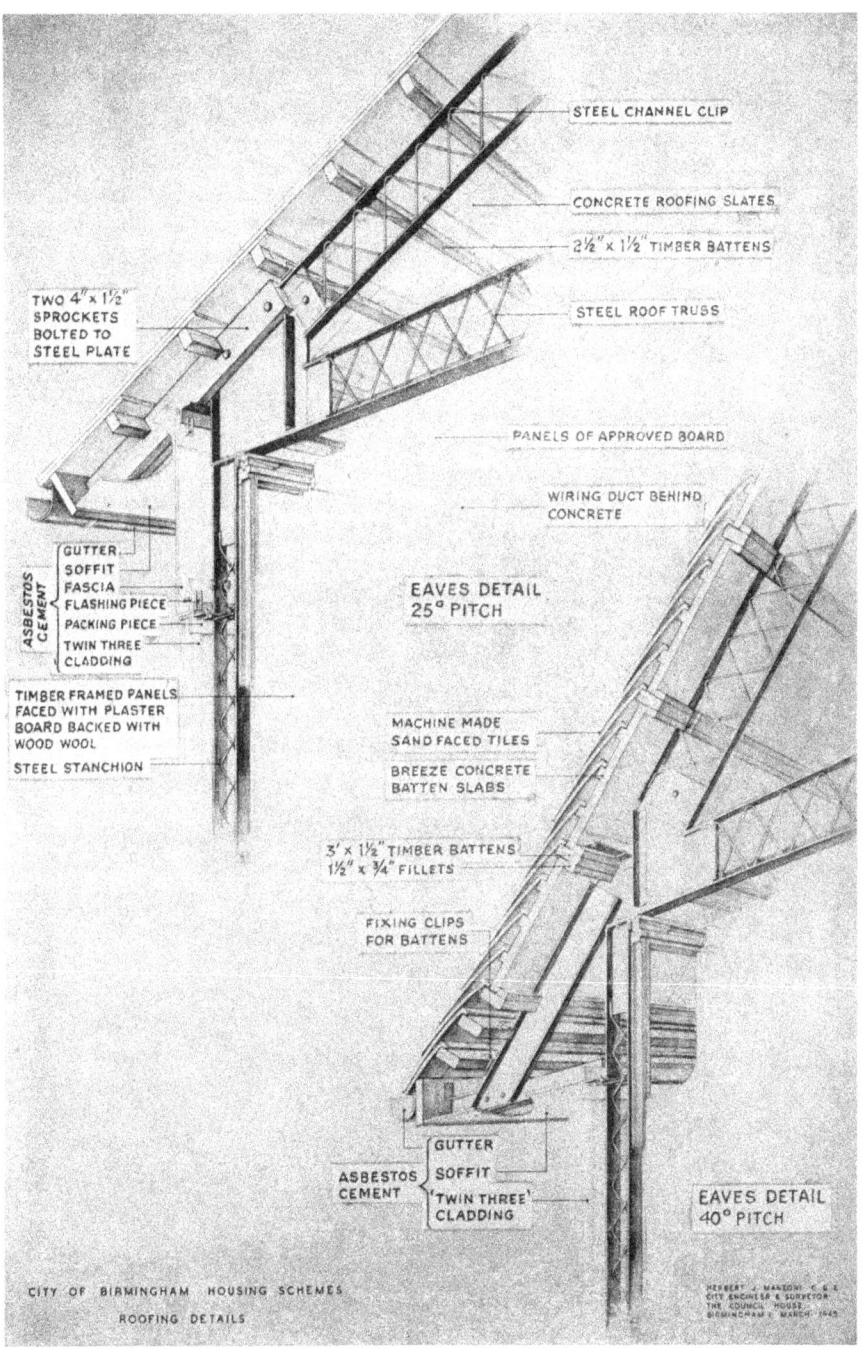

PLATE 20

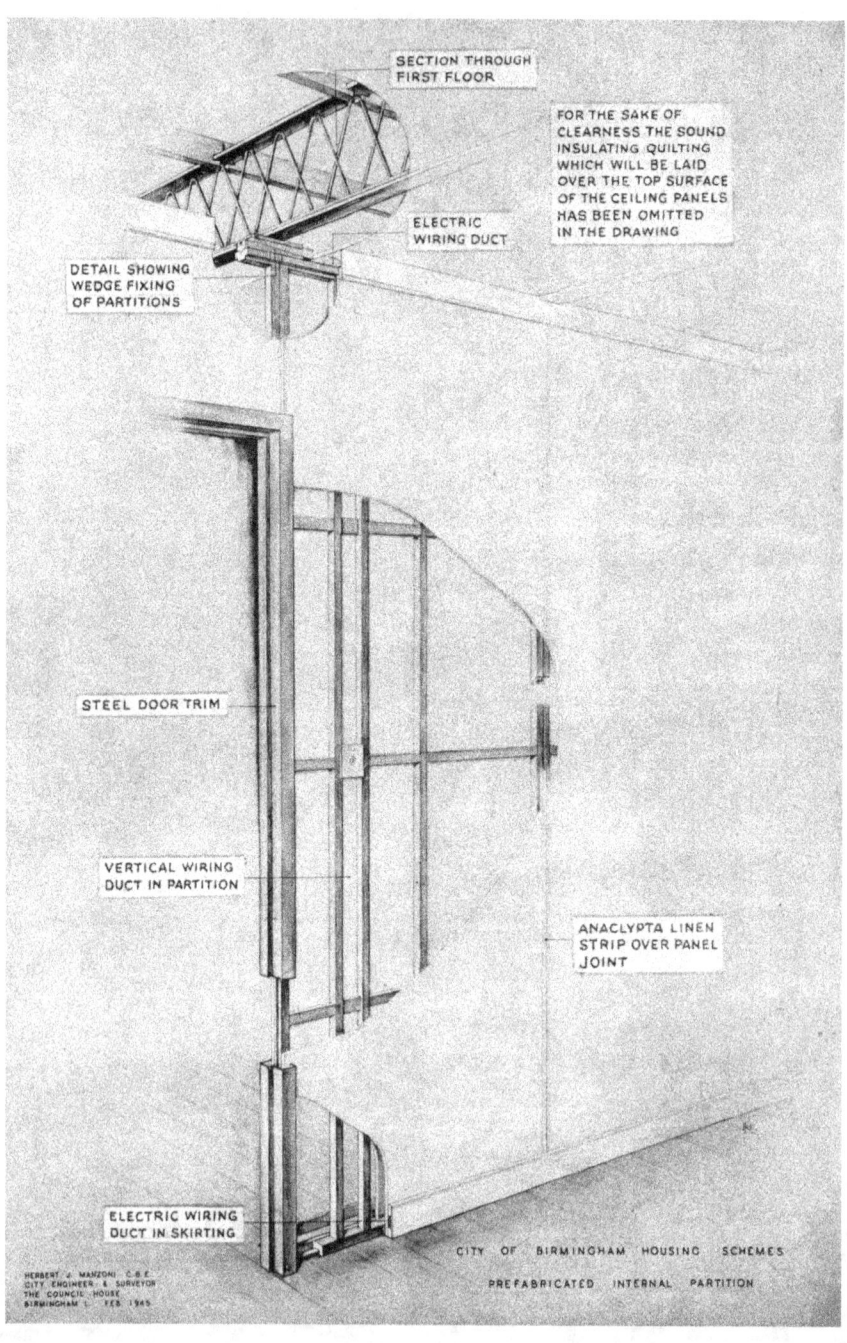

PLATE 21

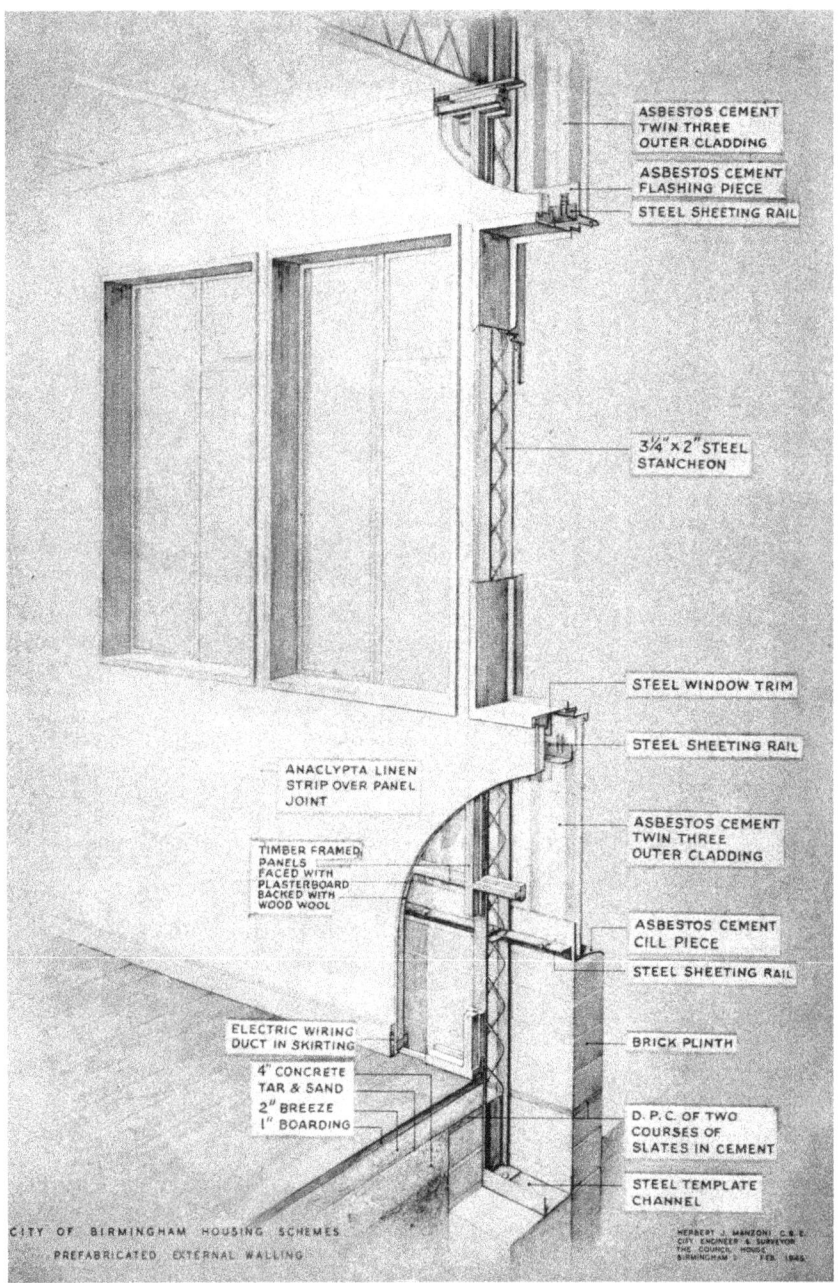

PLATE 22

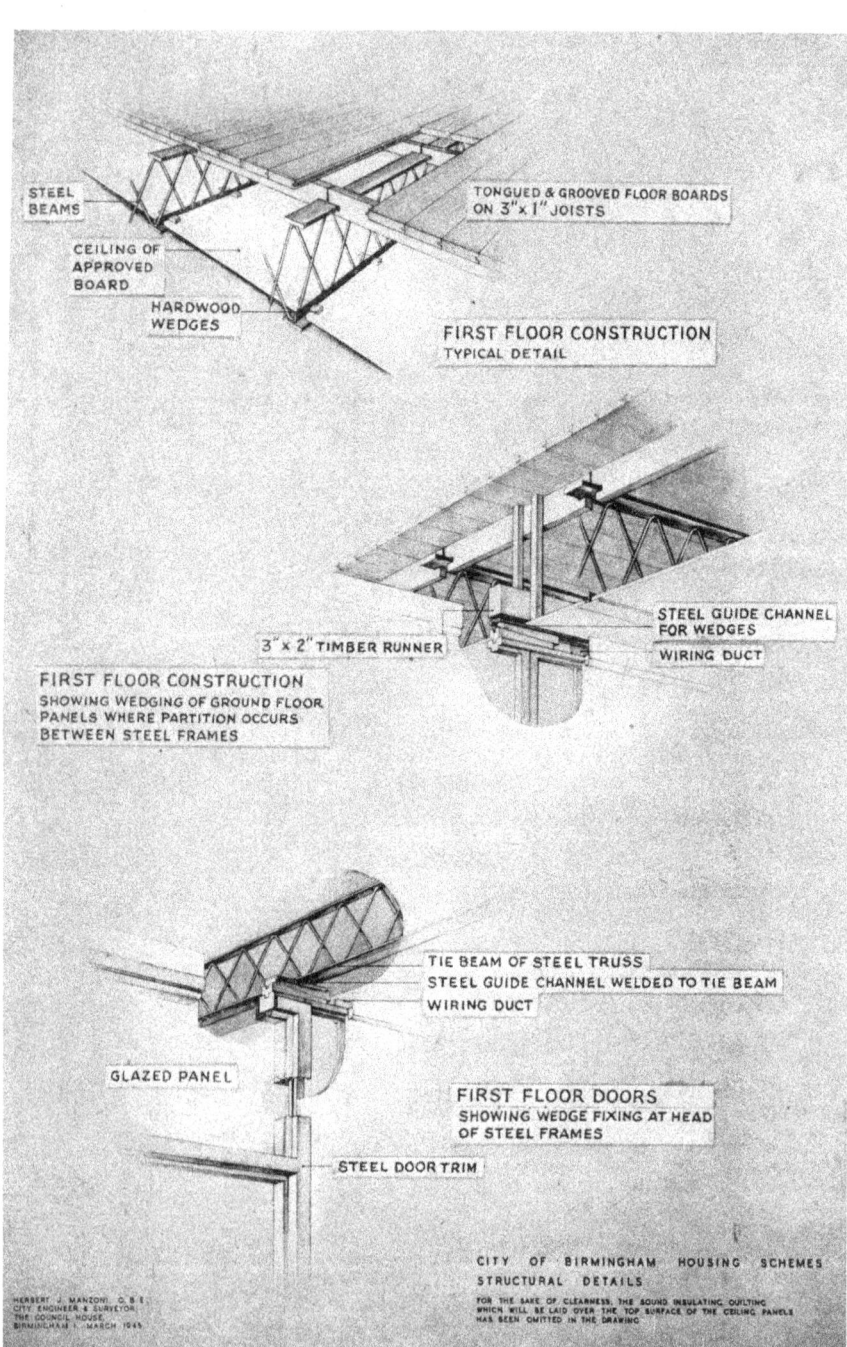

PLATE 23

Stages in erection, and complete Mark II demountable house designed by Arcon. (See page 41.)

(These photographs and those on plates 24 and 25 are reproduced by courtesy of Arcon.)

PLATE 24

To the left: The porch of the Mark IV demountable house designed by Arcon.

Below: The kitchen of the Mark IV demountable house designed by Arcon.

PLATE 25

To the right : Plumbing unit designed by Arcon.

To the left : Collapsible wringer and wash-up designed by Arcon

PLATE 26

Above : A pair of Type A Howard houses in course of erection at Datchet. Architect : Frederick Gibberd, F.R.I.B.A., A.M.P.T.I.

Below : Garden elevation.

Interiors are shown opposite, and plumbing unit on plate 28. (See pages 42 and 48.)

(*This photograph and those on plates* 27 *and* 28 *are reproduced by courtesy of John Howard & Company, Ltd.*).

PLATE 27

Two interiors of the Type A Howard house. Architect: Frederick Gibberd, F.R.I.B.A., A.M.P.T.I. To the left, the dining room, and to the right, the kitchen.
(See opposite, and plate 28.)

PLATE 28

Two-storey plumbing stack of Type A Howard house erected at Datchet, Berkshire. (See plates 26 and 27)

PLATE 29

Elevation of the third prototype houses (Type C) produced by the British Iron and Steel Federation. Architect: Frederick Gibberd, F.R.I.B.A., A.M.P.T.I.; Consulting Engineer: Donovan B. Lee, M.Inst.C.E. Permanent in construction, it has occupied 540 hours for erection as against 600 hours for the temporary Portal bungalow. This prototype was designed on a 3-ft. 9-in. grid, the panels being two storeys high, providing both structural frame and finish. The outside facing is of sheet steel finished with a paint "harl." A vitreous enamelled fillet covers the vertical joints and no screws or bolts are visible. The internal equipment, including staircase, plumbing, space heater, and kitchen fittings, have all been designed for prefabrication. (See pages 35, 45, 46 and 47.)

(The illustrations on this plate and on plates 30 *and* 31 *are reproduced by courtesy of the British Iron and Steel Federation.)*

To the left: Erecting one of the two-storey wall panels. No special tackle or cranes are needed, the panels being stood on end and pushed into position.

PLATE 30

Type C houses produced by the British Iron and Steel Federation: elevation.

Below: Porch of Type C house produced by the British Iron and Steel Federation. This is one of a variety of porches proposed, constructed of T-section steel, with sheet steel roof. Provision for growing plants is provided on one side, and weather protection on the other.

PLATE 31

Type C house, produced by the British Iron and Steel Federation : living room. (See plates 29 and 30.)

To the left: Loading wall panels on to a lorry for transport, the panels being slid into a specially designed crate.

PLATE 32

Models of a pair of Keyhouse Unibuilt houses designed by Grey Wornum and Richard Sheppard, FF.R.I.B.A. (See pages 26, 27 and 28.) (See opposite, also plates 34 and 35.)

PLATE 33

Above: First prototype Keyhouse Unibuilt houses, Coventry: front elevation.
Below: Model of a group of Keyhouse Unibuilt houses.

PLATE 34
On this plate and opposite, appear progress photographs of erection of the Keyhouse Unibuilt system prototype house.

Above: November 4, foundations.
Below: November 8, erection of ground floor chassis.

PLATE 35
Above : November 10, completion of chassis and trusses.
Below : November 16, fixing of outside cladding.
(See plates 32, 33 and 34.)

PLATE 36
Braithwaite prototype houses at Hendon, Middlesex, produced by Braithwaite & Co. Engineers, Ltd.   Architect : F. R. S. Yorke, F.R.I.B.A.
(*This photograph and those on plates 37, 38 and 39 are reproduced by courtesy of Braithwaite & Co. Engineers, Ltd.*)

PLATE 37

On this page, and on plate 38 are shown progress photographs of four stages in the erection of the Braithwaite prototype houses. (See also opposite page and plate 39.)

PLATE 38
(See plates 36, 37 and 39.)

PLATE 39

To the left: Plumbing unit in the Braithwaite prototype house.

To the right: Kitchen in the Braithwaite prototype house: refrigerator, cooker, built-in cupboards and washing machine are included.

PLATE 40

Side elevation of the Jicwood stressed skin bungalow. Architect: Richard Sheppard, F.R.I.B.A.

On the opposite plate are shown :—
  Above : Corner of living room in Jicwood bungalow.
  Below : Bedroom in the Jicwood bungalow.
  (See pages 56 to 60.)

(*These three illustrations are reproduced by courtesy of Jicwood Limited.*)

PLATE 41

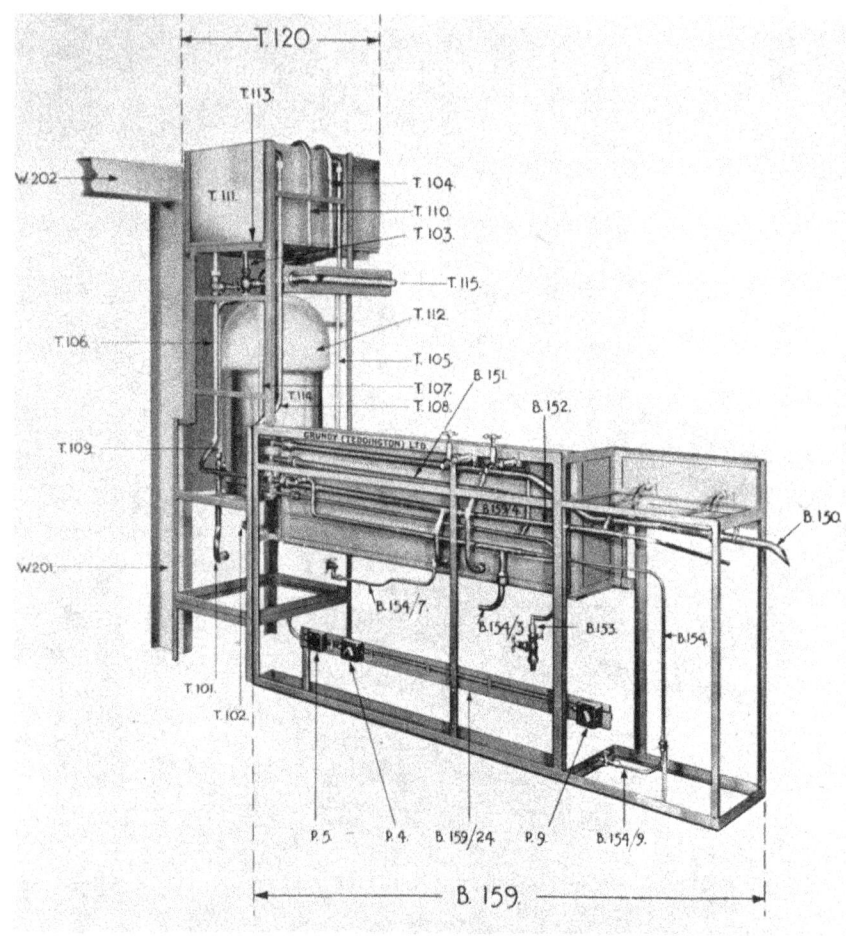

PLATE 42

Prefabricated plumbing unit designed and manufactured by Grundy (Teddington), Ltd., for the Ministry of Works Engineering Dept., for use in the Ministry of Works emergency houses. Key to the numbered parts appears opposite.

(*Reproduced by courtesy of Grundy (Teddington), Ltd.*)

Key to numbered parts of Grundy prefabricated plumbing unit shown on plate 42.

T.120	*Angle iron tower unit.*
T.101	Primary return cylinder to boiler.
T.102	Primary low from boiler.
T.103	Cold water feed to cylinder.
T.104	Vent to primary flow.
T.105	Mains cold water supply to cistern.
T.106	Cold water draw-off.
T.107	Hot water draw-off.
T.108	Overflow.
T.109	Return from towel rail.
T.110	Hot water supply vent.
T.111	Cold water cistern.
T.112	Glass insulated mattress.
T.113	Cork mattress.
T.114	Copper hot water cylinder.
T.115	Chromium plated towel rail.
B.159	*Angle iron basin unit.*
B.150	Overflow.
B.151	Hot water draw-off.
B.152	Cold water draw-off.
B.153	Mains cold water supply to cistern.
B.154	Gas service.
B.154/3	Union to gas meter.
B.154/7	Union to wash boiler.
B.159/4	Removable access panel over bath.
B.159/24	Metal channel to carry electric cable.
W.201	Warm air shroud for stove.
W.202	Warm air ducting
P.4	Pyrotenax cable, connection for wash boiler.
P.5	Pyrotenax cable, terminal box for cooker.
P.9	Pyrotenax cable, connection for refrigerator.

PLATE 43

Model of premiated design for timber house designed by John P. Tingay, A.R.I.B.A., for the Timber Development Association. Ground and first floor plans appear on plates 44 and 45.

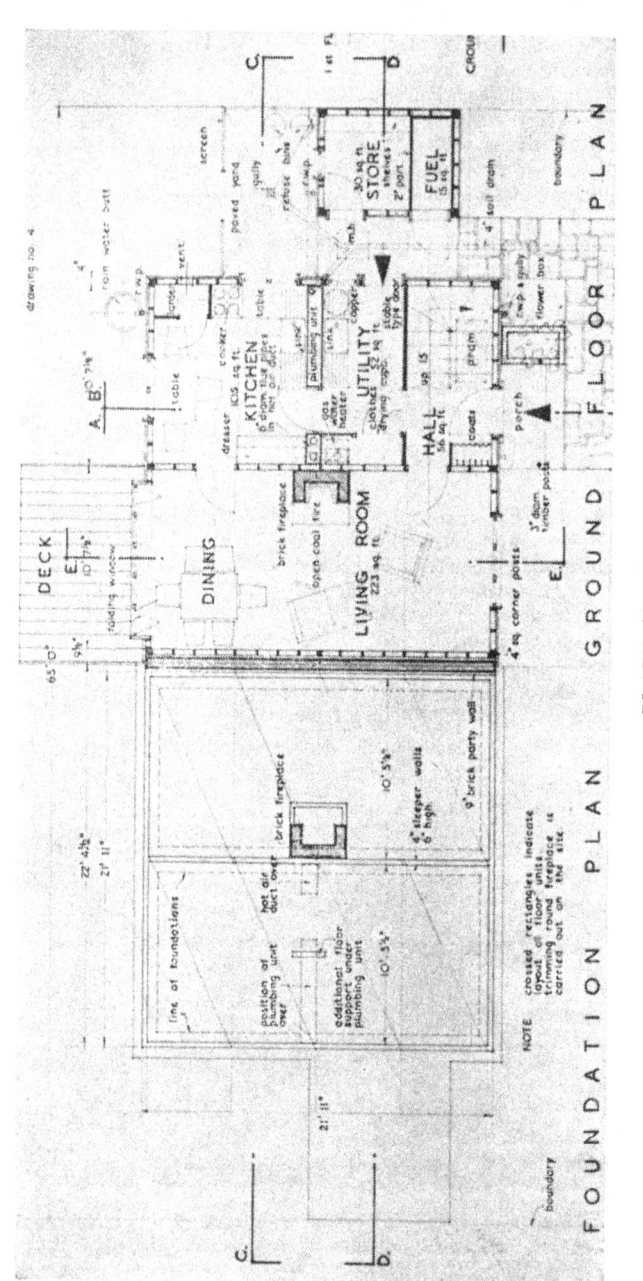

PLATE 44

Ground floor plan of a timber house designed by John P. Tingay, A.R.I.B.A. First floor plan appears opposite. (See plate 43.)

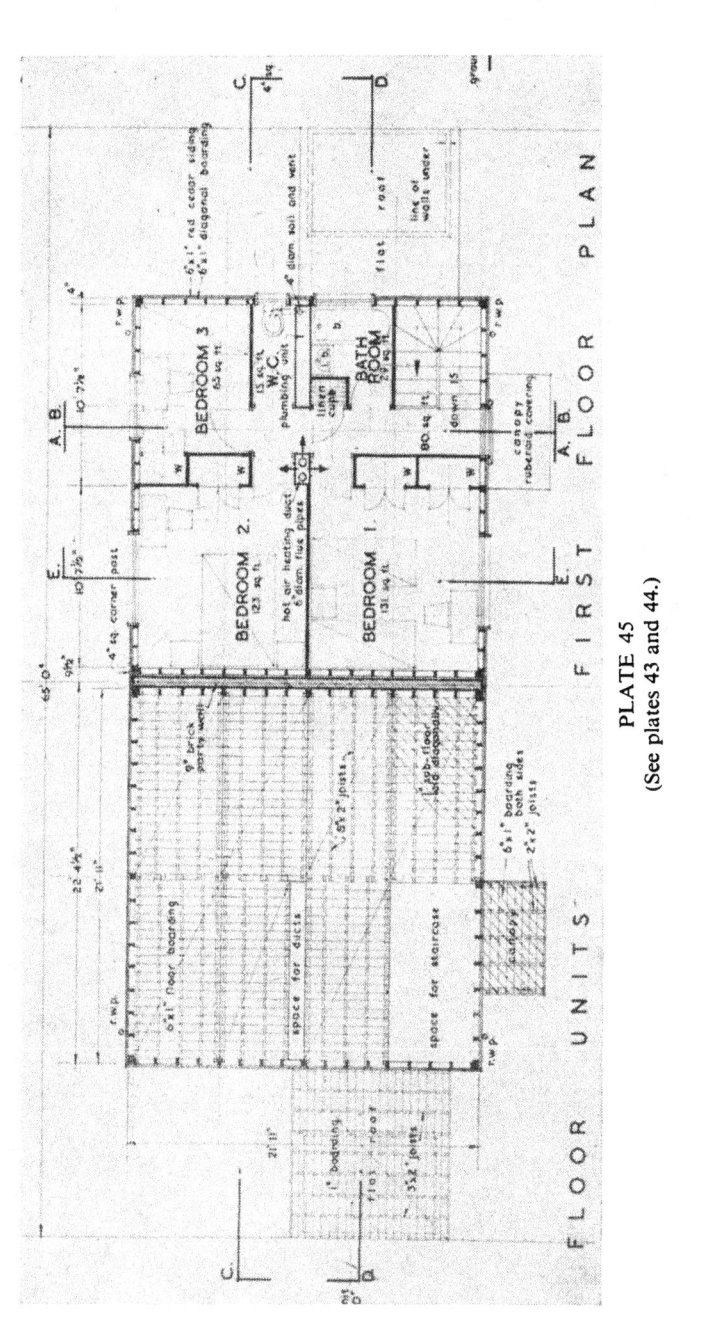

PLATE 45
(See plates 43 and 44.)

PLATE 46

To the left: The bathroom side of a gas package kitchen, designed by Jane Drew, F.R.I.B.A., and produced by Littlewoods Ltd., of Liverpool.

Below: The kitchen side of a gas package kitchen, produced by Littlewoods Ltd., of Liverpool.

PLATE 47
Part of the kitchen unit, including cooking equipment, sink and draining board, for a gas package kitchen, designed by Jane Drew F.R.I.B.A., and produced by Littlewoods Ltd., of Liverpool.

PLATE 48

Part of the Denham prefabricated plumbing unit which uses a one-pipe system.
(*Reproduced by courtesy of W. N. Froy & Sons, Ltd.*)

For Product Safety Concerns and Information please contact our EU representative GPSR@taylorandfrancis.com
Taylor & Francis Verlag GmbH, Kaufingerstraße 24, 80331 München, Germany

www.ingramcontent.com/pod-product-compliance
Lightning Source LLC
Chambersburg PA
CBHW061446300426
44114CB00014B/1863